Training and Assessing Non-Technical Skills

A Practical Guide

Training and Assessing Non-Technical Skills

A Practical Guide

By

Matthew J. W. Thomas

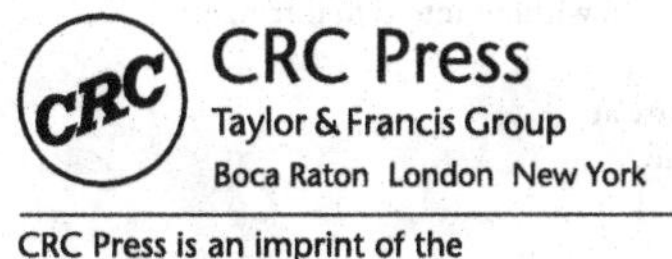

CRC Press is an imprint of the
Taylor & Francis Group, an informa business

CRC Press
Taylor & Francis Group
6000 Broken Sound Parkway NW, Suite 300
Boca Raton, FL 33487-2742

Printed on acid-free paper

International Standard Book Number-13: 978-1-4094-3633-1 (Paperback)

Visit the Taylor & Francis Web site at
http://www.taylorandfrancis.com

and the CRC Press Web site at
http://www.crcpress.com

Contents

Preface

This book has been written for anyone involved in the development or delivery of non-technical skills training programs. Specifically, it is aimed at those who work in any high-risk industry, where safety and productivity are critical. It has been designed to provide practical guidance, drawn from the scientific literature, which can assist in creating training programs that work.

While the term *non-technical skills* is relatively new, the concept has been integral to professional practice in many work domains for centuries. The term describes much of what we previously might have called *craftsmanship*, *seamanship* or *airmanship*. For instance, it has always been well accepted that to be a safe and successful mariner on the high seas meant much more than to be able to deploy sails and rigging and steer the ship in the right direction. The term *seamanship* attempted to capture the essence of expertise beyond the simple technical ability to sail. Seamanship described an almost intangible ability to be able to interpret subtle changes in weather, to effectively coordinate crew and to decide when it might no longer be safe to continue. Today, we know that these, once almost mystical, qualities of an expert can actually be defined, specified and most importantly, developed as skills through training. We now call the essence of seamanship *non-technical skills*.

The purpose of this book is to assist non-technical skills training programs to achieve their goals of enhanced safety and productivity by ensuring that training and assessment practices are built on a solid foundation of what we know to be effective and appropriate techniques. No longer do we rely on years of experience, guided as an apprentice under a master, to informally develop these skills. Over the last few decades, organisations that undertake high-risk activities, such as commercial aviation, healthcare and nuclear power generation, have worked hand in hand with Human Factors scientists to create and validate training programs to develop non-technical skills. This book seeks to share the lessons on what constitutes an effective approach to training and assessment in this domain.

For the last few decades, I have had the good fortune to be one of these Human Factors scientists. I have had the pleasure of working with high-performing nurses, commercial airline pilots, surgeons and anaesthetists, control room operators and locomotive engineers, to name but a few. I have loved learning about what it takes to be an expert in these domains while simultaneously being in awe of what these women and men are able to achieve by way of expert performance in highly demanding and often stressful work domains. In this work, I have consistently seen how non-technical skills contribute directly to safety at the sharp end.

A very large part of what has inspired this book are the practitioners and researchers who have dedicated their time to the science of training non-technical skills and who have contributed to the scientific literature in this field. In this book, I probably refer to only a small selection of these people. However, every paper or book I cite represents a significant body of work in and of itself. This book attempts to synthesise this work, and in many respects, should be read as a lesser summary of all the great work that has already been published in this field. To this end, I have intended the book to be a practical guide, and I urge readers to use this book as a roadmap to point them in the direction of the many more detailed studies of each specific aspect of training and assessing non-technical skills that are referred to in this book.

The book has been designed in part to be a companion to the definitive text on non-technical skills, *Safety at the Sharp End.*[1] While *Safety at the Sharp End* provides an exploration of the need for non-technical skills training and examines in detail the main components of non-technical skills as they relate to safe operations, the text does not focus on the 'nuts and bolts' of designing training and assessment programs. This book aims to expand on the specifics of developing a non-technical skills training program and to provide guidance within the context of adult learning theory and the science of training and assessing skills in a vocational context.

The first part of the book provides an introduction to non-technical skills and outlines how non-technical skills contribute to enhanced safety. This part concludes with an overview of the historical development of non-technical skills training programs from the late twentieth century to now.

The second part of the book provides a detailed overview of the primary considerations that need to be addressed in designing a non-technical skills training program. This part covers topics such as adult learning theory, a set of principles to guide both training and assessment of non-technical skills, and concludes with a brief overview of putting a non-technical skills training program together, from the perspectives of instructional design and needs analysis, and the sources of evidence that can be drawn on to address organisational and individual needs through a non-technical skills training program.

The third part of the book focuses on very specific strategies for training and assessing the basic core categories of non-technical skills:

1. Situation awareness
2. Decision-making
3. Communication and teamwork
4. Task management

There are definitely many more categories of non-technical skills that are important to safe and efficient performance. However, these four domains cover a basic common core seen in many existing non-technical skills training programs. Each chapter begins with a brief summary of the domain of non-technical skills, situating it within the historic context of accidents and incidents in high-risk industries. Each chapter then addresses in detail the practicalities of training and assessment of that non-technical skills domain in the classroom, in the simulator and on the job.

The training and assessment framework presented in each chapter first presents the *core knowledge* that is required to develop and deploy non-technical skills in practice. Then, each of the *constituent skills* is discussed prior to exploring the practicalities of training and assessment in the simulator and on the job.

The book closes with a vision for the future enhancement of non-technical skills training programs and sets out some areas where our journey to continually enhance human performance in high-risk industries may continue.

1. Flin, R. H., O'Connor, P., and Crichton, M. (2008). *Safety at the Sharp End: A Guide to Non-Technical Skills*. Aldershot, UK: Ashgate Publishing.

Acknowledgements

The ideas presented in this book reflect an enormous amount of work undertaken by practitioners and researchers in the development of programs for training and assessing non-technical skills across a wide range of industry contexts. Writing this book was easy in terms of drawing on this work, but also hard in deciding just what to present from the existing literature. Our field of non-technical skills is full of impassioned people dedicated to enhancing safety and efficiency in some of the most challenging forms of human endeavour. I apologise to any of my colleagues whose contributions to the field I may have not adequately represented.

In writing this book, I was assisted greatly by suggestions, along with detailed and frank reviews of draft chapters, from colleagues including Wayne Martin, Sally Ferguson, Anjum Naweed, Ganesh Balakrishnan and Robert Cannon. The book, while imperfect in many respects, is so much better for their critical reflection and suggestions. I am indebted and grateful for the time they took away from their own work to reflect on aspects of this book.

My thanks also go to my family, who gave me both the physical and the cognitive space required in the writing process. I was blessed to be able to remove myself to the farm of my parents David and Heather, or to the beautiful house of my in-laws Leray and Brian overlooking the Southern Ocean.

While I love my discipline of Human Factors and the amazing industries I am able to work with, my true love resides with my wife and daughters, Sarah, Miranda and Florence.

Author

Matthew Thomas is the deputy director of the Appleton Institute at CQUniversity and director of Westwood-Thomas Associates. He works on providing scientifically defensible solutions to enhancing human performance and managing safety in high-risk industries. His research expertise lies with human error, non-technical skills, incident investigation and data use within safety management systems.

He completed his PhD in virtual learning environments in the year 2000, and since then has explored innovative solutions to a wide range of Human Factors issues in high-risk industries. He has provided solutions to enhance the non-technical skills of professionals working in a cross-section of high-risk industries, including commercial aviation, rail, maritime, healthcare, defence, mining, utilities and others.

He contributes to a group of not-for-profit organisations worldwide and currently is the chair of the Council of the Australian Patient Safety Foundation, the immediate past president of the Australian Aviation Psychology Association and a member of the Australian Advisory Board of the Flight Safety Foundation.

1

Non-Technical Skills

A Primer

A Story of Revolutions and the Advent of High-Risk Industries

The story of human endeavour is one defined by truly remarkable achievements, worthy of celebration. From relatively humble beginnings as a species of hunter-gatherers roaming the savannah of Africa, we evolved and travelled from Africa into Eurasia and beyond.

Only 10,000 years ago, the *Neolithic Revolution* saw a transition from our traditional hunter-gatherer existence to a more stable social structure. This was enabled through our development of agriculture, through the ability to domesticate certain crops and animals, which in turn saw the creation of permanent settlements and the advent of many of the social structures we still see today. From small settlements grew larger villages, which in turn interacted through commerce. Then, larger social structures were created, and onwards to larger civilisations. Empires rose and fell as humans became truly dominant over the rest of the natural world.

Then, only 250 years ago, the *Industrial Revolution* saw another huge transition in human society and the emergence of a modern world. From a world where almost everything was created by hand, now machines dominated manufacture. Prior to the Industrial Revolution, we, as a species, were reliant on simply harnessing the power of natural elements: earth, air, fire and water. The risks associated with these elements had been known for much of our evolution as a species. Accordingly, the management of those risks was second nature to us as a species. Safety was a matter of respecting the elements, which were knowable and observable. As cultures developed, mythology was created to provide cautionary tales and describe the parameters of human capability. In ancient Greece, the legend of Daedalus and Icarus sets this out precisely. After crafting a set of wings from wax and feathers, Daedalus warns Icarus to respect the technology and not to commit an act of hubris by attempting to do what only gods, not mortals, could achieve. Icarus, however, flew too close to the sun, melting the wax, whereby his wings were destroyed and he fell to his death in the sea. In the pre-industrial world, everyday risks were much more universally understood.

In contrast to the agrarian world, the Industrial Revolution created a new class of high-risk human enterprises. Factories were born, and mass production was achieved through the use of machine tools. As never before, new technologies enabled us to transform the elements in ways that extended human capability far beyond what previously could have been imagined. What was once considered hubris was now considered innovation and growth.

As a defining component of the Industrial Revolution, the steam engine radically changed both industry and, in turn, transport. The advent of the steam engine quickly enabled mass rapid transport, and within decades, railways were commonplace. These new capabilities for the transportation of people and goods also led, across time, to significant social change.[1]

Only 150 years ago, the potential of electricity was realised, and the *Second Industrial Revolution* occurred. The period leading up to World War I was one of extremely rapid technological advancement fuelled by the potential provided by electricity. As never before, humans could harness the power of the physical world and put it to use in serving our natural tendencies towards creativity and innovation. Across successive stages of industrial revolution, exponential growth in technology occurred, and the previously unimaginable become commonplace. In a tiny space of time from an evolutionary perspective, we had created a world entirely reliant on high-risk industries, a term that describes any endeavour where a breakdown in safety can lead to a catastrophic outcome in terms of loss of life or significant environmental and economic damage.

Much has been written about the emergence of high-risk industries and the ways in which we have attempted to manage the risks associated with these forms of enterprise.[2,3] The definition of a high-risk industry is largely self-evident, and the term captures all human endeavours where there is potential for significant harm to people or the environment. Through these new industrial imperatives, human society created space for endeavours where risk was often unimaginable until it appeared, usually with catastrophic outcomes.

Today, the high-risk industry takes many forms, a reflection of the way in which our remarkable achievements as a species have come with myriad new ways of causing catastrophic harm to ourselves and our environment. From catastrophes in the generation of power, such as Chernobyl and Three Mile Island, through to accidents in processing industries, such as Bhopal, the new ways in which we interact with elements from the earth can have catastrophic consequences. Likewise, the miracles of flight have been associated with catastrophic events, as is seen every month in commercial aviation and less frequently in disasters such as the Challenger and Columbia space shuttles. Even in the healthcare industry, where we have achieved remarkable feats in the face of disease, patients are frequently harmed or killed by unmanaged risk.

The scientific domain of Human Factors has formed an important element of improving safety in high-risk industries. Human Factors is the science of people at work; it deals with applying an understanding of our characteristics, strengths and weaknesses to the design of our tools, technologies and systems of work. One of the most critical considerations of a sophisticated approach to Human Factors is the need to frame our daily endeavours in high-risk industries within the context of our extremely rapid transition from pre-agricultural hunter gathering to a post-industrial technological world. From an engineering perspective, each and every human design feature reflects so much more of the millions of years of our pre-technological hunter-gatherer ancestors than it does of our nascent exposure to high-risk industries. We are infants in a complex play-space, one that we have only just created ourselves, and it is up to us to ensure that we do ourselves no harm. Unfortunately, the way in which we have traditionally approached safety management has not always served us sufficiently well in this regard.

Safety at the 'Sharp End' of Operations

In their definitive text, Rhona Flin, Paul O'Connor and Margaret Crichton describe how non-technical skills are essential to maintaining safety at the 'sharp end' of operations in high-risk industries. By the term 'sharp end', they refer to the individuals and teams who undertake the tasks that have the greatest difficulty and the highest degree of risk associated with them. Across all human endeavours, the safety of day-to-day operations resides to a large degree in the hands of these skilled operators. Every day, safety is *created* by the actions of pilots, surgeons, train drivers, process control room operators and mechanics, to name but a few. For the vast majority of the time, the overall safety and productivity of a large and complex industrial system are achieved by their skilful operation. However, in the rare instance when something goes wrong, inevitably it is the operators at the sharp end who come under immediate scrutiny. However, the focus on non-technical skills as a critical component of the management of risk has not always been in place.

Across most modern industries, safety at the sharp end was traditionally constructed on three main foundations: (1) competent people; (2) standard operating procedures; and (3) appropriate equipment.[4] These foundations are designed to ensure consistency in the way in which work is performed, and organisational *confidence* that work will be performed in a safe and efficient manner, and are often referred to as the *Work Process Model*, or *Nertney's wheel*.

Technical Competence: Regardless of the nature of work, technical competence is of critical importance. The control room operator is trained in the technical operation of the Supervisory Control and Data Acquisition (SCADA) system, the commercial pilot is trained in the use of aircraft automation, and the train driver is trained in the traction and braking systems, among many other parts of their job. These forms of training are often competency-based in nature and follow a strict training curriculum leading to an assessment of competency or proficiency in a specific task.[5] Once deemed competent, the operator is formally certified through a licence, ticket or rating. Assessment of technical competence ensures that a person can perform the requisite tasks to get the job done.

Standard Operating Procedures: In the modern world, nearly every technical task performed is guided by a set of standard operating procedures, which generally provide a step-by-step sequence of sub-tasks that dictate how that task is to be performed. Standard operating procedures may be developed by the original equipment manufacturer, by a professional body, or by the organisation in which the work is to be performed. The specification of standard operating procedures ensures that a person consistently performs the task as planned.

Appropriate Equipment: The technical design of the on-going maintenance and reliability of equipment is the third part of the foundation for safe work in industrial settings. Ensuring the equipment is fit for purpose and reliable against breakdown and malfunction has always been an important focus of activity. Over the last century, considerations from a Human Factors perspective have also been important, such that the interface between people and equipment is designed in a manner that is sensitive to the strengths and weaknesses of both human and machine.[6]

There is a classic axiom that states: 'workers typically do what the boss *inspects* rather than *expects*'. Therefore, as a mechanism to defend against the natural human tendency to innovate, forms of worker oversight have evolved in most high-risk work environments. For many years, oversight was seen primarily as the role of government to provide regulation and inspection as a formal mechanism to ensure safety. From the first Factory Act in the UK, passed in 1883, an inspectorate was mobilised to protect the interests of workers. Over a century later, and only after catastrophic events, the emphasis shifted from government regulation to the active involvement of an organisation in managing its own risks.[7] This shift was driven largely by the Robens Committee, which recommended that an organisation's management must assume responsibility for the organisational management of risk. This recommendation, and its embodiment in 1974 within the UK Health and Safety at Work Act, set out a philosophy of so-called 'self-regulation'. Whether self-regulation takes the form of supervision on a construction site, a regular simulator-based proficiency check, or audits performed by the safety department, these mechanisms are designed to ensure that technical competence is being maintained and standard operating procedures are being followed.

This traditional approach to safety management has been referred to as the 'boots, belts, and buckles' approach, emphasising regulation and the technical protection of the workforce.[8] Organisations across the globe still invest heavily in these three areas to ensure safety at the sharp end. However, history tells us that these three foundations alone are not sufficient to ensure that work is always performed in a safe and efficient manner.

CASE STUDY: SAFETY ON THE CONSTRUCTION SITE

John attended a state-of-the-art construction training school, where brand new simulators were used to develop skills across a common set of machines used in the construction industry. As part of this training, John was certified as competent in the operation of large excavators and given his ticket to operate these machines on work-sites. As a result of his extensive experience in gaming, he was able to achieve a high degree of precision in positioning and operating the bucket. He was technically competent.

On his second week on site, John was tasked with using the excavator to clear a trench for new services. He studied the work plan and made note of the location of a gas main that ran between the area to be cleared and the footings of an adjacent building. He diligently followed the pre-start checklist and ensured that everything was safe for the job. He followed the standard operating procedures.

As John was a new operator, the site foreman decided to monitor his work and provide any assistance if necessary. He was given a high degree of supervision for the job. John proceeded with the excavation, but when he was halfway through the job, he identified that the actual location of the trench and the footings of the adjacent building did not entirely match the engineering drawings he had seen, which specified the location of the gas main. He queried this with the site foreman, who assured him the gas main had been located as safely outside the pegged area for the trench. John was uneasy, but he was urged by the site foreman just to get on with the job.

The next minute, John felt the bucket hit something hard, and a sudden rush of high-pressure gas made it evident that he had ruptured the gas main.

The organisation had set up the three main foundations of safety: (1) technical competence; (2) standard operating procedures; and (3) supervision. However, these were insufficient to ensure safety.

What was missing was a focus on *non-technical skills*. If John had been trained in the techniques of *assertiveness*, and had put these into practice, he would have stopped work until the position of the gas main was verified.

Unfortunately, when something goes wrong, operators at the sharp end are all too often judged by their technical performance and whether or not they adhered to standard operating procedures. This has led to the generic estimation that at least 80% of incidents and accidents are attributable to the human at the sharp end, and more often than not, dismissed as yet another 'human error'.[9]

A Stupid Human Broke My Machine

The focus on human error as the primary cause of incidents and accidents has always raised the ire of the Human Factors specialist and the safety scientist. This mono-dimensional explanation of a failure of a complex sociotechnical system of work, while convenient, is overly simplistic. As Dekker[10] suggests, this approach can lead to the 'bad apple' model of safety management, whereby safety is enhanced simply by 'removing' any individuals who have been involved in incidents or accidents.

CASE STUDY: REMOVING A BAD APPLE

Jane was a new engineering graduate and found her first job working for a large multinational mining company on an open-cut coal mine site. On the day of the incident, she was driving a light vehicle to pick up colleagues who had been performing survey work at a remote section of the mine site.

It was a standard operating procedure at this site to make positive communication with all heavy haul trucks when in a light vehicle crossing haul roads. Jane followed this procedure, but due to her unfamiliarity with the site, she stated she was ready to cross number 3 haul road, when in fact she was at the intersection of number 5 haul road.

When she had been given the all clear to cross, she proceeded into the intersection. Immediately, she saw that she was about to drive into the path of a haul truck, so she swerved out of the road. Unfortunately, her evasive manoeuvre was insufficient, and the 300 tonne haul truck hit and crushed the rear section of her light vehicle. Jane was shaken, but not physically injured.

After being tested for the presence of drugs or alcohol, she was sent home from work. The next day, she was required to meet with the site manager. The site manager stated that after investigation, it was found that she had broken one of the 'non-negotiable' safety policies for the site. As a consequence, there was no alternative except instant dismissal.

However, simply focussing the investigation on what rules had been breached failed to understand the root causes of the incident from the perspective of non-technical skills. Had a proper investigation been undertaken, it would have revealed that Jane had lost situation awareness and mistaken her position on the haul roads due to confusing signage and distractions from excessive non-essential radio communications. Getting rid of a 'bad apple' often ignores the ways in which human factors contribute to incidents, leaving the system vulnerable to repeat occurrences. Including an analysis of non-technical aspects of performance is a critical part of any incident investigation and may identify areas where skill development is needed.

The history of safety science in the last few decades has challenged this focus on human error as the primary cause of accidents and incidents. This has been done primarily by arguing for an approach that emphasises the organisational and systemic failures that create the conditions in which an accident can occur. The well-known author James Reason[11,12] describes two contrasting models of accident analysis: (1) the *person* approach; and (2) the *systems* approach.

Person Approach: This approach focuses on the unsafe acts of individuals. This model constructs human error as the primary cause of incidents and accidents. This approach focuses on the sharp end of operations.

Systems Approach: This approach accepts that humans are fallible and errors are to be expected. This model constructs incidents as a result of poorly designed systems that allow error to breach the defences. This model actively shifts the focus away from the sharp end of operations.

The purposeful shift away from the person approach in favour of the systems approach has been a necessary reaction to the simplistic focus on human error and individual culpability that has marred the history of safety management in the industrial world. However, this very systems-centric approach may have inadvertently worked against the realisation of non-technical skills as another critical foundation of safety at the sharp end.

A Basic Introduction to Non-Technical Skills

Non-technical skills are the *cognitive* and *social* skills that complement workers' technical skills.[13] As Rhona Flin and her colleagues define in their definitive book on the topic *Safety at the Sharp End*, non-technical skills are 'the cognitive, social and personal resource skills that complement technical skills, and contribute to safe and efficient task performance'.[14]

In many respects, non-technical skills can be seen as the 'glue' that holds together operations, and the 'enablers' of safety and efficiency. As this book will argue, they are so important that they should be considered as a fourth pillar of operational performance at the sharp end, alongside technical performance, standard operating procedures and safety assurance activities.

Many frameworks have been developed that identify the basic domains of non-technical skills, but there is no one definitive list of

Table 1.1 Core Domains of Non-Technical Skills

CORE SOCIAL SKILLS	CORE COGNITIVE SKILLS
Communication	Situation awareness
Leadership	Decision-making
Teamwork	Task and workload management

all the non-technical skills required to ensure safe and efficient operations. Indeed, the specific broad domains of non-technical skills are likely to be quite different depending on the industry context and the nature of the work being performed. However, based on several decades of research and development in the area, a common core of basic non-technical skills can be set out. Table 1.1 outlines the core domains of non-technical skills.

CASE STUDY: NON-TECHNICAL SKILLS AND A ROAD TRAFFIC TRAUMA PATIENT

From the second a road traffic trauma patient arrives in the emergency department, their care involves the coordinated activity of a large multidisciplinary team.

The first stage involves handover from the first responder paramedics to hospital staff. This is a complex process of *communication* and *situation awareness*, where the hospital trauma team must quickly develop a shared mental model of the patient's current status and injuries.

If the patient is unstable, effective *communication* and *teamwork* are required to provide advanced life support activities in the resuscitation room. Once the patient is stabilised, complex *decision-making* activities must occur, often involving complex diagnostic interchanges between emergency physician, radiologist, pathologist and surgeon.

While each of the individuals involved has very specific technical expertise, which is quite different from that of other members of the team, the ultimate outcome for the patient relies on coordinated multidisciplinary work. This coordination is almost exclusively reliant on the non-technical skills of each member of the multidisciplinary team.

The Role of Non-Technical Skills in Day-to-Day Operations

Non-technical skills encompass a wide range of skills that are all critical to maintaining safe performance in the high-risk work environment. But how do these skills actually lead to safe and efficient operations?

First, they can be seen to be essential to the effective execution of planned performance.[15] Non-technical skills play an important role in good practice, and while not always specified, are typically present in any workplace setting. For instance, the communication process of effectively sharing information between members of the team is a fundamental aspect of team performance.

Perhaps one simple way of describing non-technical skills in day-to-day operations is to examine the performance of elite sporting teams. An elite football team is made up of a group of individuals with technical prowess, strength, agility and stamina. These are the technical aspects of the skill-sets they bring to the match. However, for those skills to be effectively translated into winning performance, another allied set of skills needs to be employed. The team members need to be able to *communicate* effectively such that they can achieve coherent game play. Individuals need to maintain good *situation awareness* with respect to the position of team members from the opposing team, such that they can open up opportunities to push the ball into attack. Individuals need to exhibit good *decision-making* under time pressure with respect to where to pass the ball and when to strike.

Each of these non-technical skills is an enabler. That is to say, in day-to-day settings, it is non-technical skills that enable the effective utilisation of technical skills and knowledge. Non-technical skills are the vehicle through which technical skills and knowledge can be applied. In some respects, they are the *art* in contrast to the technical *science* of performance.

The Role of Non-Technical Skills in Responding to Operational Complexity

Beyond the role non-technical skills have in effective execution of day-to-day work, they also play a critical role in responding to situations where the messy complexities of the real world present challenges to the operators at the sharp end.

One way of describing this messy complexity of everyday operations is through the use of the threat and error management (TEM) model. Developed by the late Professor Robert Helmreich and his colleagues at the University of Texas Human Factors Research Project, the TEM model describes how safe and efficient performance is maintained in the less than perfect world of day-to-day operations.[16–18] The basic premise of the TEM model is that day-to-day operations do not always occur in a simple and benign environment in which the standard operating procedures as described in the operational manual can be executed. Rather, the context of day-to-day operations is messy and complex, and safe and efficient operations are maintained by operators effectively responding to and managing this complexity. In an elegantly simple way of describing this complexity, the TEM model categorises three aspects of the messy day-to-day world that need to be the focus of specific actions to manage them: (1) threats; (2) errors; and (3) undesired states. Recent research has demonstrated that effective management of these aspects of the messy day-to-day world is made possible through the deployment of non-technical skills.[19]

First, *threats* are defined as factors that increase the likelihood of errors and include aspects such as adverse environmental conditions, system malfunctions and the errors of other parties.[20] Maintaining safe and efficient performance involves identifying these threats as they arise and taking action to mitigate the risks associated with them. Here, non-technical skills such as *vigilance*, *problem identification* and *task prioritisation* are critical.

Second, *errors* are described as action or inaction by the operators at the sharp end that leads to deviations from expected performance.[21] Much has been written about human error, and it is now accepted that human error is a natural, unavoidable and often predictable part of any human endeavour. Errors are typically classified as (1) *lapses*, which are memory-based failures; (2) *slips*, which are errors in the execution of a task; (3) *mistakes*, which are errors in decision-making or judgement; and (4) violations, which are wilful deviations from procedures or rules.[12] Again, maintaining safe and efficient performance involves detecting any errors as they occur and taking action to return the system to its usual state. While different forms of error are associated with different types of mitigation, non-technical skills such as *planning and preparation*, *monitoring and cross-checking*, and *evaluation of plans* are important aspects of error detection and management.

Finally, *undesired states* are defined as the outcome of poor TEM, where the safety of the operation has been allowed to be compromised. Situations such as maintaining flight below minimum safe altitude, or the undetected deterioration of a patient, are typical examples of undesired states arising from poor TEM. Again, non-technical skills such as *assertiveness* and *leadership* are pivotal in ensuring these states are detected and recovery is achieved.[22]

The well-known author James Reason has recently introduced the notion of 'heroic recoveries', whereby individuals and teams are able to recover from extreme undesired states where an operation has teetered on the brink of catastrophe. The notion of the heroic recovery highlights how a range of non-technical skills contribute to restoring safety from the brink, and the human skills of complex decision-making, workload management and teamwork are critical ingredients in these recoveries.

As we examine each domain of non-technical skill throughout this book, we will focus on the way in which the skills contribute to the effective management of the complex and dynamic world of operations in high-risk industries.

The Relationship between Non-Technical Skills and Safety

Non-technical skills have a two-way relationship with safety: poor non-technical performance is linked to safety being compromised, and good non-technical skill performance can be seen to enhance safety.

First, deficiencies in non-technical performance, such as poor *decision-making* or inadequate *vigilance*, have been consistently associated with incidents and accidents. In this context, the oft-cited metric that 80% of accidents are caused by human error is overly simplistic and generally misguided. Even models that suggest it is not error alone that causes accidents, but a trajectory that begins with more remote causes in the form of latent organisational failures, fail to adequately describe the role of non-technical skills in maintaining safety in the face of latent failures and error.[11,23]

It is well accepted that human error is both common and ubiquitous. Even in the ultra-safe industry of commercial aviation, where the accident rate worldwide is less than one in every million flights, there are usually many errors made by flight crew on every flight.[19] Therefore, the translation of an error rate of, say, five in every flight to an accident rate of one in every million flights suggests that there is a lot more than just human error involved in the causation of accidents. Non-technical skills make a significant contribution to the management of everyday human error. Therefore, it can be argued that it is poor non-technical performance that allows error to compromise the safety of an operation.

Conversely, high-performing operators can be defined in terms of excellent non-technical performance. To this end, non-technical skills enhance safety, or using more recent terminology, non-technical skills contribute to the *resilience* of high-risk industries.

CASE STUDY: THE UNSTABLE APPROACH

The morning flight was uneventful, and in clear skies the crew began their descent towards their destination. As was the norm, the crew had anticipated being cleared to approach the airfield from out over the water, after flying over the hills to the east of the city. The captain, as pilot flying for this sector, adjusted his descent profile with this plan in mind.

When the aircraft was approximately 20 miles from the field, Air Traffic Control (ATC) offered the crew a 'straight in' approach, as the wind at the field was light and variable. The first officer asked the captain what he thought about a 'straight in' approach, as it would leave them significantly high and fast on the descent profile. The captain simply said it would be fine, 'no problems', so the first officer confirmed with ATC that they would accept the 'straight in' approach.

The aircraft was vectored by ATC towards the runway, and with 10 miles to run, the aircraft was still high on the approach profile. The first officer saw that the captain, in his attempts to get the aircraft back on the approach profile, was not beginning to slow the aircraft such that it could be configured for landing.

The first officer sensed that the captain had not fully anticipated the implications of accepting the 'straight in' approach.

At eight miles to run, the first officer suggested the landing gear could be lowered to increase drag and reduce speed. The captain took a moment to respond, and then called for landing gear to be lowered. The aircraft began to decelerate, and the captain called for the first stage of landing flaps to be extended. However, the first officer did not extend the flaps, as the aircraft was still too fast and above the aircraft's limit speed for the first stage of flaps. At five miles to run, the crew were able to begin configuring flaps for landing, but by two miles, the aircraft was still high and fast. The first officer suggested that they perform a 'go around' and follow the missed approach procedure, but the captain ignored this input. The captain continued the now unstable approach and attempted a landing. The aircraft bounced and floated for some distance, before finally settling on the main landing gear approximately halfway down the runway. With the runway available, the captain could not slow the aircraft sufficiently, and it continued off the end of the runway, being substantially damaged. Several passengers were injured as a result of the evacuation.

This event encapsulates where deficiencies in non-technical performance are causal factors in accidents. First, the captain exhibited poor *decision-making* in accepting the 'straight in' approach to the airfield. Further, calling for flap extension above the flap limit speed demonstrates degraded *situation awareness* through not being aware of the aircraft's actual speed. Finally, the first officer's failure to insist on a missed approach or using the standard 'emergency language' and taking over control of the aircraft demonstrates deficiencies in *communication* and *assertiveness*.

Even given the relatively short history of formal non-technical skills training programs, a body of scientific evidence now highlights the effectiveness of such training programs in enhancing the safety performance of organisations operating in high-risk industries.

Drawing direct conclusions between non-technical skills training programs and overall safety performance of an organisation is problematic. In industries such as aviation, with an already very low failure rate, there are many other variables that prevent a causal relationship being established between training and accident rates.[24] However, with our detailed understanding from incident investigations of how failures in non-technical performance are implicated directly in accident causation, we can infer that if we lift the overall non-technical performance of operators, we in turn reduce the likelihood of accidents occurring. This suggests, therefore, that we should be looking for changes in performance of individuals and teams pre-and post-training as a measure of the effectiveness of non-technical skills training programs.

Very early in the development of crew resource management in aviation, evaluation studies demonstrated positive changes in attitude and performance associated with training. For instance, even by the early 1990s, recurrent seminar-based training coupled with Line Oriented Flight Training (LOFT) in a high-fidelity simulator had been shown to result in positive changes in flight crew behaviour.[25] Since then, a considerable body of similar evidence has emerged.

More recent research has established formal relationships between the non-technical performance of flight crew and their management of error in normal commercial flight operations. High performance on a range of non-technical skills was shown to predict the crew's response to operational threats and the management of their own errors. For instance, if a junior crew member demonstrated high levels of *assertiveness*, errors were more likely to be effectively managed.[19]

In a similar study, significant relationships were observed between crew performance in the domains of communication, situation awareness and workload management and crews' response to operational errors. Further, it was demonstrated that non-technical skills were important countermeasures for ensuring safe performance in crews who had not worked together before.[26] Previously, it had been suggested that standard operating procedures were the most important component in supporting unfamiliar teams in maintaining safe operations. However, in this study, it was shown that while unfamiliar crews made significantly more errors, their non-technical skills enabled effective error management, and therefore no significant impact on safety was observed. This contributes another dimension to the body of evidence supporting the role of non-technical skills in high-risk industries.

From Craftsmanship to Non-Technical Skills

Another important way to contextualise the concept of non-technical skills is to describe them with reference to the concept of *craftsmanship*.

Throughout the history of human endeavour, the concept of the *master craftsman* has defined the peak of skilled work. Craftsmanship describes both a true mastery of skills and an engagement with the task such that a more enlightened approach can be adopted.[27] Rather than simply being able to use a set of woodworking tools to build an object, the master craftsman carpenter is able to create the object in a manner that embeds a higher degree of form and function, of beauty and practicality. The master craftsman transcends technical competence and integrates the activity of both the head and the hand in their work.

This concept of transcending technical competence is also embedded in the somewhat ill-defined terms of *seamanship* and *airmanship*. Together, these terms describe the qualities of an operator who is able to augment their technical skills with a greater understanding of the nuances of the environment in which the work takes place. The holder of seamanship is able to use their experience to anticipate complex effects of wind against tide. Likewise, the holder of airmanship is able to draw on their advanced knowledge of aerodynamics and aircraft systems to create a novel solution to a problem not previously encountered.

Traditionally, it has been thought that the master craftsman and the artisan, as well as those who show high degrees of seamanship and airmanship, have developed these extra elements of performance either as a natural gift, or through many tireless hours of time and effort; that is to say, through a combination of aptitude and expertise. While, indeed, this may be true, the theory of non-technical skills would suggest this is not the only path to higher levels of performance.

Take, for example, another form of master craftsman, in this case a doctor who demonstrates an exceptional bedside manner and remarkable diagnostic insights. If we adopt the perspective of craftsmanship and its parallel pathways of aptitude and expertise, it is easy to explain the high levels of performance exhibited by this professional. However, this approach ignores to a large degree the role of formal training in the development of the doctor's skills in communication (bedside manner) and clinical decision-making (diagnostic insights).

The concept of non-technical skills presents a set of challenges to our traditional thinking about high-performing workers. First, the domain of non-technical skills challenges the conception that high levels of performance in areas such as communication are derived from pre-existing aptitude. This 'nature versus nurture' argument is an important consideration. Many people make the assumption that these areas of performance are innate or come as pre-developed skills in the adult worker. This type of thinking assumes that the ability to communicate effectively has been developed by schooling and parenting prior to an individual entering the workforce. Aspects of performance such as conflict resolution and assertiveness are developed through youth and early adulthood on the sports field, between friends and family, and through the many awkward moments of transition into adulthood. However, in contrast to this position, the domain of non-technical skills holds that skills such as communication are *not* simply the product of pre-existing aptitude, but rather, are skills like any other, which can be developed through training and subjected to competency-based assessment.

Second, the domain of non-technical skills challenges the concept that expert performance is the result of exposure to a wide range of experiences in the work environment. Rather, the domain of non-technical skills holds that well-designed scenario-based training can provide a better environment in which to develop many of the aspects of high performance that have traditionally been associated with the notion that an expert is one who has accumulated more than the arbitrary number of 10,000 hours, or 10 years, of work experience.

The traditional approach to non-technical skills, which largely ignores any form of formal training, can still be seen in the way in which many organisations approach recruitment. Selection processes often focus on an applicant having the requisite technical competence, as demonstrated through certification, and appropriate attitudes and patterns of interpersonal interaction, as demonstrated through psychometric testing. Indeed, many organisations screen to ensure that these cognitive and social skills are already in place through rigorous selection processes that include team activities and scenario-based interviews.[28]

While this line of thought is wonderfully humanistic, it largely precludes the possibility that safety-critical cognitive and social skills can be developed and enhanced by training and assessment.

Open and Closed Skills: Soft and Hard Skills

In the context of this book, which is written primarily for high-risk industries, we will adopt the terminology of non-technical skills to refer to those broader cognitive and social skills that complement technical skills in successful job execution. However, in a wide range of other industries, the general concept of non-technical skills is used, either using the same terminology or using other terms such as 'soft skills' or 'open skills'.

Across a range of workplace settings, the division between closed and open skills has been used to differentiate the types of skills to be developed and the types of training interventions that can be used to support this skill development.[29] In this context, technical skills are referred to as *closed*, because they describe skills that are performed in a standardised manner in specific contexts. These skills are often deployed against a backdrop of rules governing how a task shall be completed. On the other hand, open skills are less prescribed and instead demonstrate a degree of adaptability to operational circumstances.

This concept of open and closed skills maps directly onto our concept of technical and non-technical skills. However, for the sake of consistency, we will use the term *non-technical skills*, although the literature relating to open skills remains entirely relevant.

The Assumptions of This Book

Given the preceding discussion, this book comes with a set of underlying assumptions, which are best described as follows.

First, it is assumed that non-technical skills are not inherent predispositions derived from personality or aptitude. Rather, it is assumed that competence in each domain of non-technical skill can be developed through purposeful training.

Second, it is assumed that non-technical skills do not effectively evolve organically as a component of expertise. Rather, it is again assumed that just as in the training of technical skills, non-technical skills can be trained from an *ab initio* level, and continually rehearsed and enhanced like any other skill.

Third, it is an underlying assumption of this book that non-technical skills form essential elements of competency, and without formal training and assessment, an operator cannot be deemed to have all the skills required to perform a safety-related role in a high-risk industry.

Together, the overall assumption is that non-technical skills *must* be effectively trained and assessed as a critical component of maintaining safe and efficient operations in high-risk industries.

References

1. Kellett, J. R. (2007). *The Impact of Railways on Victorian Cities*. Oxon, UK: Routledge.
2. Shrivastava, P., Mitroff, I. I., Miller, D., and Miclani, A. (1988). Understanding industrial crises. *Journal of Management Studies, 25*(4), 285–303.
3. Perrow, C. (2011). *Normal Accidents: Living with High Risk Technologies*. Princeton, NJ: Princeton University Press.
4. Nertney, R. J., and Bullock, M. G. (1976). *Human Factors and Design*. Springfield, VA: U.S. Energy Research and Development Administration.
5. Leung, W.-C., and Diwaker, V. (2002). Competency based medical training. *British Medical Journal, 325*(7366), 693.
6. Fitts, P. M. (1951). *Human Engineering for an Effective Air Navigation and Traffic Control System*. Washington, DC: National Research Council.
7. Hale, A. R., and Hovden, J. (1998). Management and culture: The third age of safety. A review of approaches to organizational aspects of safety, health and environment. In A. Feyer and A. Williamson (Eds.), *Occupational Injury, Risk, Prevention and Intervention* (pp. 129–165). London, UK: Taylor & Francis.
8. Thomas, M. J. W. (2012). *A Systematic Review of the Effectiveness of Safety Management Systems*. Canberra, Australia: Australian Transport Safety Bureau.
9. Reason, J. (1990). *Human Error*. Cambridge, UK: Cambridge University Press.
10. Dekker, S. W. A. (2006). *The Field Guild to Understanding Human Error*. Aldershot, UK: Ashgate Publishing.
11. Reason, J. (1997). *Managing the Risks of Organizational Accidents*. Aldershot, UK: Ashgate Publishing.

12. Reason, J. (2000). Human error: Models and management. *British Medical Journal, 320*, 768–770.
13. Flin, R., Martin, L., Goeters, K.-M., Hormann, H.-J., Amalberti, R., Valot, C., and Nijhuis, H. (2005). Development of the NOTECHS (Non-Technical Skills) System for assessing pilot's CRM skills. In D. Harris and H. C. Muir (Eds.), *Contemporary Issues in Human Factors and Aviation Safety* (pp. 133–154). Aldershot, UK: Ashgate Publishing.
14. Flin, R. H., O'Connor, P., and Crichton, M. (2008). *Safety at the Sharp End: A Guide to Non-Technical Skills*. Aldershot, UK: Ashgate Publishing.
15. Fletcher, G., McGeorge, P., Flin, R. H., Glavin, R. J., and Maran, N. J. (2002). The role of non-technical skills in anaesthesia: A review of current literature. *British Journal of Anaesthesia, 88*(3), 418–429.
16. Helmreich, R. L. (2000). On error management: Lessons from aviation. *British Medical Journal, 320*, 781–785.
17. Helmreich, R. L., Klinect, J. R., and Wilhelm, J. A. (1999). Models of threat, error and CRM in flight operations. In R. S. Jensen (Ed.), *Proceedings of the Tenth International Symposium on Aviation Psychology* (pp. 677–682). Columbus, OH: Ohio State University.
18. Helmreich, R. L., and Merritt, A. C. (2000). Safety and error management: The role of crew resource management. In B. J. Hayward and A. R. Lowe (Eds.), *Aviation Resource Management* (pp. 107–119). Aldershot, UK: Ashgate Publishing.
19. Thomas, M. J. W. (2004). Predictors of threat and error management: Identification of core non-technical skills and implications for training systems design. *International Journal of Aviation Psychology, 14*(2), 207–231.
20. Helmreich, R. L., Wilhelm, J. A., Klinect, J. R., and Merrit, A. C. (2001). Culture, error and crew resource management. In E. Salas, C. A. Bowers and E. Edens (Eds.), *Improving Teamwork in Organizations* (pp. 305–331). Mahwah, NJ: Lawrence Erlbaum Associates.
21. Klinect, J. R., Wilhelm, J. A., and Helmreich, R. L. (1999). Threat and error management: Data from line operations safety audits. In R. S. Jensen (Ed.), *Proceedings of the Tenth International Symposium on Aviation Psychology* (pp. 683–688). Columbus, OH: Ohio State University.
22. Helmreich, R. L., and Merrit, A. C. (1998). *Error and error management*, Technical report 98-03. Austin, TX: University of Texas Aerospace Crew Research Project.
23. Shappell, S. A., and Wiegmann, D. A. (2001). Applying reason: The human factors analysis and classification system (HFACS). *Human Factors and Aerospace Safety, 1*(1), 59–86.
24. Helmreich, R. L., Chidester, T. R., Foushee, H. C., Gregorich, S. E., and Wilhelm, J. A. (1990). How effective is cockpit resource management training? Issues in evaluating the impact of programs to enhance crew coordination. *Flight Safety Digest, 9*(5), 1–17.
25. Helmreich, R. L., and Foushee, H. C. (2010). Why CRM? Empirical and theoretical bases of human factors training. In B. G. Kanki, R. L. Helmreich and J. M. Anca (Eds.), *Crew Resource Management* (2nd edn., pp. 3–58). San Diego, CA: Academic Press.

26. Thomas, M. J. W., and Petrilli, R. M. (2006). Crew familiarity: Operational experience, non-technical performance, and error management. *Aviation, Space and Environmental Medicine, 77*(1), 41–45.
27. Sennett, R. (2008). *The Craftsman*. New Haven, CT: Yale University Press.
28. O'Connor, P., and Cohn, J. (2010). *Human Performance Enhancement in High-Risk Environments*. Santa Barbara, CA: Praeger Security International.
29. Yelon, S. L., and Ford, J. K. (1999). Pursuing a multidimensional view of transfer. *Performance Improvement Quarterly, 12*(3), 58–78.

2
The Classroom Is Dead
Long Live CRM ...

The Need for Training Non-Technical Skills

During the nineteenth and twentieth centuries, industrial advancement followed a common cycle of rapid technological development, component or system failure, engineering solutions, and then a short period of stability before further technological advancement. In many industrial contexts, lessons were learnt the hard way. System failures were catastrophic in nature, and were cruel to both the human and the environment alike.

In many industries, the period between the 1950s and the 1970s saw rapid change with the increasing availability of automation and technological control. Similarly, these decades saw the first human landing on the moon, the advent of the video cassette recorder (VCR) and the Walkman, the pocket calculator and the first personal computers. These were the decades of modern technology. The 1950s were also the birth of the so-called 'jet age', when modern jet transport became not only a reality, but even commonplace for some. However, this birth was not without complications. Engineering and technical issues plagued the early years of modern jet transport, and a long series of aviation accidents ensued. During the 1960s, in the early days of commercial jet aviation, fatal accident rates were at an annual average of over 10 accidents per million departures worldwide. The exemplar of this technical unreliability was the infamous de Havilland DH 106 Comet, the very first jet transport aircraft.

The de Havilland Comet was the first jetliner to be placed into commercial service, and in its first year, it carried tens of thousands of passengers. However, after several accidents when early models of the aircraft failed to become airborne, a significant engineering redesign of the wing was undertaken. A few years later, a number of Comet aircraft catastrophically broke up mid-flight. After extensive investigations, it was discovered that the corners of the square windows were suffering stress fractures after repeated cycles of normal pressurisation and depressurisation.[1] Again, engineering studies and redesign were called for to address the underlying failures of new technology.

For commercial aviation, the majority of these engineering and technical issues had been resolved by the early 1970s, and the modern commercial jet airliner was both safe and reliable. During the 1970s, the international accident rate in commercial aviation dropped to an average of approximately three accidents per million departures, from the 10 per million departures in the 1960s.[2] The jet era brought rapid advances in technology, and in turn, safety. However, accidents still continued to occur. The spectre of so-called 'pilot error' became the focus of attention, and again, aviation safety was no longer a domain primarily for engineers and their technical solutions.

This was not the first time in the history of aviation human factors that the engineers of complex systems were able to sit back and point their accusing fingers at the areas of selection, training and operations. More specifically, individual error and culpability became the easy substitution for technical failures. Quickly, the safety philosophy of 'a stupid human broke my machine' became entrenched, and 'pilot error' was frequently identified as the root cause of catastrophic accidents. The technological advances in equipment safety revealed the substrate of natural human performance variability, and also deficiencies in pilot training and assessment.

While our language of system safety has come a long way since this simplistic 'single cause' model of human error, during this time the finger was pointed squarely at the human operator as the primary cause of accidents. It is of interest that this increasing focus on the individual operators and their errors came within a decade known as 'the me decade'. According to commentators such as Tom Wolfe, the 1970s were a decade that saw increasing individualism. Dubbed 'the *Me* decade', the 1970s shifted the focus from community and family to the individual.[3] It could be argued that this increase in individualism and autonomy also resulted in an increase in individual accountability and culpability. Hence, human error and turning to ways to improve human performance on the flight deck became the focus for enhancing safety in commercial aviation.

While this philosophy of individual culpability remains alive and well in many industries, it did not sit well with the desire to create an ultra-safe system of commercial aviation. To this end, the question 'What can we do to address these areas of vulnerability and fallibility?' was quickly raised. The engineering fraternity were limited in their ability to respond to this question, so industry turned to those whose endeavour was the science of being human. They turned to psychology, and they turned to research.

Research into Pilot Error

During the 1970s, NASA undertook a series of investigations into the emergent problem of entirely avoidable accidents caused by pilot error. One of these studies was a large-scale simulator study of flight crew performance undertaken at the Ames Research Center and led by visiting Royal Air Force physician and pilot H. P. Ruffell Smith.

For perhaps the first time, a research study identified that much of the variability observed in pilot performance was associated with deficiencies in the way in which crews utilised all the resources available to them. The study demonstrated the relationship between error and factors such as workload management, communication and decision-making.[4]

The Ames pilot error study concluded that flight crew performance, and especially the performance of captains, could be enhanced through the development of specialised training programs that focused on the effective utilisation of all resources available to them, including the human resources of other crew members (team performance) as well as the use of technical and other information to support their operations (situation awareness and leadership).

This study set out the basic framework for what today has become known as non-technical skills training and assessment. However, the journey to our modern training programs was only just beginning.

Industry-Driven Development

While NASA was conducting these research studies, industry did not simply sit back and watch from the sidelines. The 1970s were a decade of critical reflection and innovation with respect to pilot training. Particular focus was placed, for perhaps the first time, beyond technical proficiency in pilot selection in the commercial aviation environment. A new focus saw the development of comprehensive command upgrade training for co-pilots to transition into the role of captain.[5] Airlines had reflected on the emergent theme of 'human error' in accidents and, quite sensibly, had started to address this issue with a focus on the characteristics of a good commander. Alongside efforts in aviation, during this period there were also considerable activities in other domains where human error had been singled out, such as the maritime industry.[6]

In Europe, KLM instituted human factors training for flight crew, drawing on the now well-known software, hardware, environment, liveware (SHELL) model of system safety. Similarly, Pan Am undertook a critical review of its training and operations manuals and created a new focus on crew interaction and crew performance.[7] By the end of the 1970s, there was sufficient momentum for industry, academia and the regulator to come together to share experiences and define best practice in training programs designed to address the spectre of 'human error'.

Resource Management on the Flight Deck: The Workshop

The results of the NASA studies, and the renewed focus on flight crew human factors issues, led to a joint NASA and industry workshop in 1979 entitled *Resource management on the flight deck*.[5] The workshop sought ways to improve safety through exploring strategies that might enhance individual and crew performance, and thus minimise human error.

The outcomes of this workshop echoed research findings of the 1970s, which suggested that the human factors issues involved in accident causation were characterised by failures in both cognitive and social skills, rather than a lack of technical ability of the flight crew. The synthesis of these research findings identified the main human factor causes of air crashes as the failures of *communication*, *crew coordination*, *decision-making* and *leadership*.[8] These were to become the very focus of the first iterations of training and assessment of non-technical skills in the form of cockpit resource management training.

The NASA workshop was the culmination of a decade of research exploring the role of pilot error in aviation accidents, and sought to lay out a framework for interventions to reduce the accident rate. The interventions focussed predominantly on training as a route towards wide-scale attitudinal and behavioural change.

Across three days in June 1979, a formidable program saw a series of presentations, first by those who had been involved in research exploring the theoretical underpinnings of human error as it manifested itself on the flight deck. Then, a series of presentations from industry champions, specifically training captains and training managers from across the globe, involved these leaders sharing their developments from the last decade. Finally, a series of interactive workshops took place mapping the potential future of applied human factors training.

At this workshop, the term *cockpit resource management* (CRM) was applied to the training of crews in making use of all available resources such that error could be avoided. The workshop was a watershed moment for the future development of programs of training and assessment of non-technical skills.

In the decade that followed, airlines continued to develop and refine their CRM training programs. The NASA workshop had clarified the focus on aspects of communication, leadership and decision-making, and airlines invested their efforts towards new ways of training in these areas.

It All Began in a Classroom ...

The origins of CRM can be found almost exclusively in the classroom, with a specific emphasis on awareness raising and knowledge development seminars. A typical initial CRM course consisted of several days in the classroom, with instructional techniques limited to lectures, case studies, videos of accident re-enactments and role-play exercises.[8,9]

The initial Federal Aviation Administration (FAA) advisory circulars on CRM training emphasised the role of what was notably referred to as 'indoctrination/awareness' training. The implication of the way in which early CRM training was conceptualised was that once flight crew were exposed to the doctrine of CRM, there would be attitudinal change, which then would result in enhanced crew behaviour on the flight deck.[10]

In 1986, NASA hosted another workshop in this area, the *Cockpit Resource Management Training Workshop*. This time, the workshop was much more focussed on sharing experiences of airlines across the globe in developing their CRM programs. Airlines such as United, Continental, Japan Air Lines, Trans Australia Airlines, US Air and Air New Zealand all presented their programs, as well as examples from the military context. At this time, there was much greater synergy between the practitioner and the scientist, and the proceedings show a general sense of excitement at developing innovative programs that could have a very real benefit to flight safety.

This brief history of the birth of non-technical skills training and assessment programs is a celebration of one industry's collective response to an emergent safety issue. The early evolution of CRM in such a brief period is the true beginning of current non-technical skills programs today, and the degree to which programs today reflect these beginnings is a testament to the work of those early pioneers.

A decade since the initial NASA workshop, the FAA published the first advisory circular on the topic: AC 120-51 Cockpit Resource Management.[11] This document provided guidance on the structure and content of CRM programs, and for the first time the regulator formally recognised the importance of non-technical skills training programs.

By the 1990s, the results of the first efforts to evaluate the tangible effects of crew resource management on actual flight crew performance began to be published. The results of a joint NASA/University of Texas project led by the late Professor Bob Helmreich, published in 1990, revealed that participation in formal CRM training was associated with a higher percentage of crews being rated positively by expert observers in both full mission simulation and observations of everyday flights.[12]

Hot Tub Therapy and Boomerangs

Early forms of CRM training were not always the great success that they had the potential to realise. For some, they were overly 'psychological' in nature, and rather than providing real practical implications for the task of flying an aeroplane, they were dismissed by some as being nothing more than 'hot tub therapy'.

This rejection of early CRM programs is of significant importance to our quest to design the very best non-technical skills training programs. Several key factors were identified that led to success or failure of the early programs. First, while the classroom-based seminar was the overarching delivery platform for CRM training, the degree to which training concepts were made relevant to the actual task of flying a commercial airliner was essential. Pilots were often critical of role-playing and teamwork games that bore little resemblance to their real-world jobs.[13] Second, there was the suggestion that CRM was not for everyone, and that particular personality traits might well be associated with a rejection of CRM concepts by individuals. Regardless, those who rejected CRM training were viewed as 'boomerangs' – they kept coming back exactly the same.[14,15]

The Escape from the Classroom Seminar

Even from the outset, the role of practical forms of CRM training had been flagged, but the practicalities were that a classroom-based seminar was the most readily achievable way of introducing what were quite new concepts to the pilot group. Even within this context, however, the importance of scenario-based training, along with the potential use of simulation in CRM training, had always featured in the initial conceptualisation of CRM. However, in reality, the first CRM programs in aviation were all about the classroom, and as a consequence, the real benefits in terms of the development of non-technical skills were not easily realised.

By now, there was some consistency in the general syllabus and curriculum for CRM programs, and more importantly, a new focus on the need for CRM programs to focus on skill development and rehearsal in the actual setting of aircraft operations. One of the true pioneers of CRM, the late Professor Bob Helmreich, stated at an early CRM workshop:

> With regard to power and activity, the trainee needs to be personally involved and actively participating in the process. The classic

> lecture / text instructional format does not provide the involvement and personal learning necessary to effect real change. In this context, LOFT with videotape feedback is one of the most powerful tools we have. I am convinced that CRM training without the chance for practice and self-observation that comes with LOFT will be relatively ineffective.[16]

Echoing these sentiments was Captain David Shroyer from United Airlines, who, in describing the United Airlines (UAL) approach to CRM, stated:

> At this point I think it is important to emphasize that all of the academic learning that may be accomplished or presented in any operation does not mean anything until a vehicle or training system is in place which can bring the complete process to fruition. Let me repeat—without such a vehicle to permit a flight crew to use the tools learned, it only garbages up the mind.[17]

In less than a decade, the commercial aviation industry had undertaken research to specify the aspects of flight crew performance underlying an untenable safety record, taken steps to articulate how these findings might translate into a training program, and realised that the early seminar-based approaches, which centred on indoctrination into a new way of thinking, needed much more in terms of actual skill development.

A true non-technical skills training program steps well outside the boundaries of the traditional classroom environment. While awareness of key human factors issues and an understanding of strategies to enhance individual and crew performance are critical, skill development involves experience and rehearsal.

Eventually, CRM was able to systematically extend beyond the classroom, predominantly via two pathways. The first of these pathways was through the already existing concept of Line Oriented Flight Training (LOFT). The LOFT approach provided the perfect vehicle for a more active and skills-based approach to training the core non-technical domains of the CRM program. Indeed, the precursor to LOFT was a training program developed at Northwest Airlines for coordinated crew training, which was designed to address the fact that regulatory mandated training did not include issues around crew coordination.[18]

The central driver of LOFT was to move away from the traditional use of high-fidelity simulation in aviation for the rehearsal of specific flight manoeuvres and responding to technical failures of aircraft systems. Traditionally, simulation-based training in aviation had focussed on the use of the simulator as a cheaper, more accessible and safer alternative to using the actual aircraft in training. Rather than taking a larger commercial airliner out of revenue service to re-create complex emergency situations during flight, the simulator enabled flight crew to practise these situations in safety. In a single training session, pilots could move through a long sequence of events they had to manage, such as engine failures, cargo fires and the malfunction of multiple aircraft systems. However, this type of use of the simulator left little opportunity for the development and assessment of non-technical skills.

As an adjunct to the traditional multi-event training, LOFT focused on crew performance during 'full mission' simulation scenarios. The LOFT scenario typically involves all aspects of flight: pre-flight preparation and briefing, departure, cruise, descent approach and landing. Crew are provided with all appropriate documentation for the flight, and the simulator instructor acts in the role of dispatch, air traffic control, cabin crew and any other role, such as engineering, as is required by the scenario. The participants perform as a normal crew and are left uninterrupted to complete the mission. Throughout the simulation, normal, abnormal and emergency events are introduced, and the crew must effectively manage these events.

The development of LOFT (and its derivatives) allowed for the first time consideration to be given to the CRM skill domains of teamwork and crew communication, situation awareness, problem identification and decision-making. Indeed, the current version of the FAA advisory circular states explicitly that the aim of LOFT is to improve total flight crew performance by integrating CRM and technical skills.[19]

Several instructional design philosophies are embedded within LOFT in order to enhance the development of non-technical skills, including pre-session briefing to activate and prime concepts relating to non-technical skills, realistic scenarios developed from operational needs, such as re-creating events reported through the safety management system of the airline, and facilitated debriefing, where the crew are able to critically reflect on their own performance. These aspects of instructional design have formed a critical foundation for subsequent non-technical skills training programs across many industries, and each will be explored in detail throughout this book.

The second pathway through which CRM emerged from the classroom was via the Advanced Qualification Program (AQP). In response to the 1987 crash of Northwest Airlines Flight 255, in which the flight crew failed to configure the aircraft for take-off and omitted to perform critical checklists, the National Transport Safety Board (NTSB) issued a Safety Recommendation on the subject of CRM training. The recommendation was that all commercial airlines review their flight crew training programs to ensure that they included simulator or aircraft training exercises that involved cockpit resource management and active coordination of all crew member trainees, and permitted evaluation of crew performance and adherence to those crew coordination procedures.[20] In light of this recommendation, along with the recommendations from a joint industry–government task force into flight crew member training, the FAA developed the AQP and formalised it through regulation in 1991.

The AQP was developed as a more flexible form of flight crew training. Rather than being bound to a 'one size fits all' training curriculum set out in prescriptive federal regulation, AQP enabled organisations to develop their own customised training curriculum based on analysis of organisational training needs.[21]

One of the major advances put forward by AQP was that airlines were required to *integrate* Human Factors and CRM concepts into all their training. No longer was CRM relegated to the classroom; rather, it was to be integrated with aspects of technical training.[22] The AQP sought to reduce the probability of crew-related errors by aligning training and evaluation requirements more closely with the known causes of human error. This was done through detailed tasks analyses of performance requirements for each aircraft type and operational type across each fleet of an airline. This prompted the development of a bespoke scenario-based curriculum, which was designed around individual and crew performance.

Formal Requirements for Training and Assessing Non-Technical Skills

Given the developments of the 1970s and 1980s, it was only a matter of time before aviation regulatory bodies would require the formal training and assessment of non-technical skills alongside existing requirements for training and assessing technical skills. In contrast to the initial work originating out of the US, Europe was first to develop a regulation that required flight crew to have their non-technical skills trained and assessed formally.

In response to this new regulatory requirement, a research consortium was developed to first review existing methodologies for training and assessing non-technical skills, and then create a system for use in Europe if nothing suitable already existed.[23] This research project eventually resulted in the creation of one of the first formal frameworks developed for the assessment of non-technical skills. The NOTECHS system provides us with a clear example of how we can talk about the basic structure of a non-technical skills framework.

The NOTECHS system describes a hierarchical framework of non-technical skills that are critical to safe and efficient commercial aviation operations. At the top level, there are four *categories* of non-technical skills specified in the NOTECHS system. These categories describe the broad domains of performance. Each of these categories is then sub-divided into *elements*, which describe more specific skill-sets. Table 2.1 provides a summary of the categories and elements found in NOTECHS.

Within the NOTECHS system, each of the elements is further described through the use of *behavioural markers* of both good and poor

Table 2.1 The NOTECHS Framework of Non-Technical Skills Categories and Elements

CATEGORY	ELEMENTS
Cooperation	Team-building and maintaining Considering others Supporting others Conflict solving
Leadership and managerial skills	Use of authority and assertiveness Providing and maintaining standards Planning and coordination Workload management
Situation awareness	Awareness of aircraft systems Awareness of external environment Awareness of time
Decision-making	Problem identification and diagnosis Option generation Risk assessment and option selection Outcome review

performance.[24] The term *behavioural markers* refers to a prescribed set of behaviours indicative of some aspect of performance.[25] Table 2.2 provides an example of behavioural markers – for the NOTECHS element *use of authority and assertiveness*.

This three-level description will be used throughout this book as the fundamental structure of a basic framework that can be deployed for the training and assessment of non-technical skills. However, we will adopt a broader set of terminology for the three levels. First, the top level will be referred to as the *domain* of non-technical skill. The second level will be referred to as the set of *constituent skills* that make up the domain. In turn, each of these constituent skills will then be described in more detail using a set of *behavioural markers* for good and poor practice. A more detailed examination of this hierarchical structure will be provided in Chapter 5, when we explore the assessment of non-technical skills.

Joint Training and Team Training

One of the first important developments in CRM training in the aviation industry was the realisation that it should not be limited to flight crew only. The tragic Air Ontario Flight 1363 that crashed on take-off from Dryden, Ontario highlighted that cabin crew, and their interaction with flight crew, form a critical element of safety. This accident involved cabin crew not conveying to the flight crew the concerns of a travelling pilot about the degree of snow and ice accumulation on the aircraft's wings. This snow and ice subsequently led to the aircraft not being able to attain sufficient altitude at the end of the runway. After

Table 2.2 An Example of Behavioural Markers

USE OF AUTHORITY AND ASSERTIVENESS	
GOOD PRACTICE	POOR PRACTICE
Takes initiative to ensure crew involvement and task completion	Hinders or withholds crew involvement
Takes command if situation requires, advocates own position	Passive, does not show initiative for decisions, own position not recognisable
Reflects on suggestions of others	Ignores suggestions of others
Motivates crew by appreciation and coaches when necessary	Does not show appreciation for the crew, coaches very little or too much

this accident, many airlines developed joint CRM training programs, in which both flight crew and cabin crew come together for training.

This innovation paved the way for multidisciplinary team training as CRM training programs became popular in many other high-risk industries.

Parallel Developments: Anaesthesia Crisis Resource Management

While the story of CRM as told so far is one of the evolution of non-technical skills training in the context of commercial aviation, this story alone fails to acknowledge the innovation happening within other industries at this time. In parallel to the evolution of CRM, non-technical skills training was rapidly being developed in other contexts. Throughout the 1980s, other high-risk industries had been keenly observing the development of CRM in commercial aviation, and it was not long before the lessons learnt in aviation were being transferred to other workplaces.

One of the first adaptions of CRM was undertaken by a group of anaesthetists at the Stanford University School of Medicine, led by David Gaba and Steven Howard.[26] This group identified that resident anaesthetists did not receive any formal training in crisis management techniques, but rather, developed the requisite skills informally through observation and their own experience. This perceived deficiency led to the development of the Anaesthesia Crisis Resource Management (ACRM) training program, a program that drew heavily on CRM in aviation. The ACRM training program used a standardised structure for the development of non-technical skills. During a half-day seminar, participants were first introduced to the concept of non-technical skills before participating in a series of simulator scenarios designed to actually develop enhanced non-technical performance.

What this group of anaesthetists did was quite remarkable, and their contribution to the current forms of non-technical skills training and assessment is unfortunately sometimes overlooked. The development of CRM for the operating theatre coincided with the development of patient simulators. While simulation had become common in industries such as aviation, maritime and nuclear power generation, it had not been successfully implemented on any scale in anaesthesia.

However, the potential had been identified, and by the late 1980s, David Gaba's group at Stanford had implemented and evaluated a successful prototype. This group immediately saw the potential for simulation to enable new forms of training, including training in non-technical skills. One of the benefits of a patient simulator for anaesthesia was identified as the ability to make errors in a safe environment and to learn from those errors without any associated patient harm.[27] Additionally, simulators provided an opportunity for the development of non-technical skills to occur alongside the development of technical skills.

For aviation, simulators pre-dated CRM and were almost exclusively used for technical training and assessment. Once CRM had become an accepted form of classroom-based training, there was some disconnect in airlines between the CRM seminar and the LOFT program, with the latter drawing on high-fidelity simulation. This was not the case in healthcare, where operating theatre simulation developed hand in hand with the development of CRM programs. This co-evolution of syllabus, curriculum and educational technology resulted in something rare and wonderful.

First of all, in healthcare, non-technical skill training was integrated with technical training from the outset. In an anaesthetic crisis simulation, drug choice and dose would be discussed alongside allocation of tasks and teamwork. Patient ventilation rate was discussed alongside communication. A disconnected IV line was an issue for a situation awareness discussion.

Second, there was a palpable focus on actual non-technical skill development in a manner that was entirely *in situ* and operationally realistic. There was no need to try to transfer learning from a classroom role-play to an in-flight emergency, as was the case in the early stages of aviation CRM and LOFT.

While healthcare continued to innovate and focus on skill development, CRM in commercial aviation languished in the classroom in many airlines, being dominated by scenario-based discussion and role-play. While the syllabus evolved through consecutive generations, the curriculum remained basically the same. A quarter of a century later, it is healthcare that now provides many of the examples of best practice when it comes to the training and assessment of non-technical skills. Accordingly, the core principles of integration, operational fidelity, rehearsal, repetition and skill development will be delved into in detail throughout this book.

The Global Spread of Resource Management Programs

The adoption of crew resource management programs in anaesthesia was not the only direction of spread. Many other industries were closely watching the developments being made by NASA, the FAA and others within the airline industry. The issues faced by aviation and the focus on 'human error' were by no means unique. A range of similar human factors issues were being identified as a critical causal factor in industrial accidents worldwide, and CRM was seen to provide a potential solution. Quickly, other industries were adapting the CRM training program to their unique environments, often working with those who had been seminal players in establishing CRM programs.

The maritime domain was also an early adopter of the CRM concept, and in many respects, may have developed such training before aviation. In the 1970s, the Warshash Maritime Centre in Southampton developed the Bridge Operations and Teamwork training program, which was a simulator-based program designed to enhance teamwork on the bridge. This was certainly a very early program designed explicitly for the development of non-technical skills.[6] By 1995, the concept of Bridge Resource Management (BRM) was formalized within the International Maritime Organization (IMO) Standards of Training, Certification and Watchkeeping (STCW).[28] The bridge of a large vessel shares a number of similarities with the flight deck and the operating theatre with respect to the critical role of non-technical skills such as communication, situation awareness and decision-making, among others. While the ways these skills are deployed, and the constraints and challenges to manage, might differ considerably, forward thinkers in the maritime industry could easily see the potential benefits of developing their own form of resource management training. Further, with the increasing use of high-fidelity simulators in the training of bridge crew, the opportunities were easy to realise.

Resource management programs were not just something for high-profile work environments such as the flight deck and the operating theatre. In the 1990s, maintenance resource management (MRM) was being developed for aircraft maintenance engineers. The best-known of these is the program developed by Transport Canada, which centred around 12 established human factors issues and error-producing conditions associated with maintenance error, the so-called 'dirty dozen'.[29] This program is still being used today and has been adapted for a wide variety of other work domains as diverse as electrical utilities.

The extractive industries have also been pro-active in developing training programs that focus on non-technical skill development. Rhona Flin and her colleagues at the University of Aberdeen were leaders in this area, serving the needs of the offshore oil and gas industries.[30,31] These programs specifically focussed on skills associated with critical incident management.[8]

More recently, in Australia, a program called Associated Non-Technical Skills has been developed for the Mines Safety Advisory Council and Department of Industry in New South Wales. The program was designed to contribute to mine safety in Australia by developing non-technical skills across the industry. This program differs from others inasmuch as it focusses specifically on embedding training in the everyday work environment through directed coaching activities, thus adopting another vehicle for developing non-technical skills.[32]

The control room environment is an area where non-technical skills training programs can provide significant benefit. Recent research has emphasised specifically the use of simulation and 3D immersive environments for the development of non-technical skills such as situation awareness.[33]

As resource management training became more and more commonplace, and the focus on the development of applied human factors competencies became explicit, the term *non-technical skills* has become the more frequently used term to describe such training. Table 2.3 provides an overview of some of the industrial domains where non-technical skills training programs have been developed and evaluated.

The Future Is Bright …

For non-technical skills training programs, the future is bright. Across high-risk industries and beyond, the realisation that non-technical skills are important and can be developed through training programs

Table 2.3 Summary of High-Risk Industries Where NTS Training and Assessment Is Currently Used

INDUSTRY	SPECIFIC EMPLOYEE GROUPS
Aviation	• *ab initio*[34] • Flight crew[13] • Maintenance engineers[35] • Cabin crew[36] • Air traffic controllers[37]
Healthcare	• *ab initio*[38] • Acute medicine[39] • Anaesthetists[26] • Nurses[40] • Surgeons[41] • Intensive care physicians[42] • Cardiac arrest teams[43] • Paramedics[44] • Obstetrics[45] • Urology[46] • Trauma teams[47] • Emergency teams[48]
Maritime	• Bridge teams[49] • Pilots[50]
Rail	• Drivers[51] • Rail safety personnel[52]
Mining and extractive industries	• Offshore oil and gas well operators[53] • Control room operators[54]
Engineering	• *ab initio*[55]
Nuclear	• Control room and operational personnel[56] • Emergency response[57]
Other	• Prison management[58]

is now well established. More importantly, considerable research has explored the most effective ways in which non-technical skills can be trained and assessed.

Individual operators and organisations alike can no longer dismiss such training programs as psychobabble-laden tree-hugging jollies. Likewise, organisations can no longer see investing in the training of non-technical skills as an expensive exercise with little tangible return on investment.

Rather, as discussed in the previous section, the scientific evidence now undeniably demonstrates that non-technical skills training programs contribute to enhanced safety performance. Without these programs in place, operators cannot be seen to be competent in all aspects of safety-critical work in high-risk industries. In turn, organisations will be seen as wanting if their approach to managing the risk of catastrophic failures does not systematically include training programs for relevant non-technical skills.

Looking further into the future, with continued scientific endeavours, theoretical development continues to take place with respect to each of the core domains of non-technical skill and the ways in which they can be better trained.

While some industries still utilise a predominantly seminar-based format, the gold standard has left the classroom behind in favour of simulation- and workplace-based programs that focus on practical skill development.

For all intents and purposes, the actual concepts of technical and non-technical skills present a divide that does not exist in practice. Safe and efficient team performance is a gestalt of a range of knowledge, skills and attitudes. However, the separate focus on non-technical skills was a necessary evil such that better definition and some momentum for dedicated training in these areas could be realised. The concept of *integrated* skills development will be picked up as a theme throughout this book. However, for the foreseeable future at least, we will still need to abstractly differentiate the technical and non-technical competencies.

References

1. Walker, T., and Henderson, S. (2000). *The First Jet Airliner: The Story of the De Havilland Comet*. Newcastle upon Tyne: Scoval.
2. Boeing Commercial Airplanes. (2011). *Statistical Summary of Commercial Jet Airplane Accidents Worldwide Operations 1959–2010*. Seattle, WA: Boeing Commercial Airplanes.
3. Wolfe, T. (1976). The 'me' decade and the third great awakening. *New York Magazine, August,* 26–40.
4. Ruffell Smith, H. P. (1979). *A Simulator Study of the Interaction of Pilot Workload with Errors, Vigilance, and Decisions*. NASA Ames Research Center Moffett Field, California: National Aeronautics and Space Administration.
5. Cooper, G. E., White, M. D., and Lauber, J. K. (1980). *Resource Management on the Flight Deck Proceedings of a NASA/Industry Workshop (NASA CP-2120).* Washington, DC: NASA.
6. Hayward, B. J., and Lowe, A. R. (2010). The migration of crew resource management training. In B. G. Kanki, R. L. Helmreich and J. M. Anca (Eds.), *Crew Resource Management* (2nd edn., pp. 317–344). San Diego, CA: Academic Press.

7. Helmreich, R. L., and Foushee, H. C. (2010). Why CRM? Empirical and theoretical bases of human factors training. In B. G. Kanki, R. L. Helmreich and J. M. Anca (Eds.), *Crew Resource Management* (2nd edn., pp. 3–58). San Diego, CA: Academic Press.
8. Flin, R., O'Connor, P., and Mearns, K. (2002). Crew resource management: Improving team work in high reliability industries. *Team Performance Management*, *8*(3/4), 68–78.
9. O'Connor, P., Campbell, J., Newon, J., Melton, J., Salas, E., and Wilson, K. A. (2008). Crew resource management training effectiveness: A meta-analysis and some critical needs. *International Journal of Aviation Psychology*, *18*(4), 353–368.
10. Federal Aviation Administration. (1993). *Advisory Circular 120-51A: Crew Resource Management Training*. Washington, DC: US Department of Transportation.
11. Federal Aviation Administration. (1989). *Advisory Circular 120-51: Crew Resource Management Training*. Washington, DC: US Department of Transportation.
12. Helmreich, R. L., Wilhelm, J. A., Gregorich, S. E., and Chidester, T. R. (1990). Preliminary results from the evaluation of cockpit resource management training: Performance ratings of flightcrews. *Aviation, Space, and Environmental Medicine*, *61*(6), 576–579.
13. Helmreich, R. L., Merritt, A. C., and Wilhelm, J. A. (1999). The evolution of crew resource management training in commercial aviation. *International Journal of Aviation Psychology*, *9*(1), 19–32.
14. Chidester, T. R., Helmreich, R. L., Gregorich, S. E., and Geis, C. E. (1991). Pilot personality and crew coordination: Implications for training and selection. *International Journal of Aviation Psychology*, *1*(1), 25–44.
15. Helmreich, R. L., and Wilhelm, J. A. (1991). Outcomes of crew resource management training. *International Journal of Aviation Psychology*, *1*(4), 287–300.
16. Helmreich, R. L. (1987). Theory underlying CRM training: Psychological issues in flight crew performance and crew coordination. In H. W. Orlady and H. C. Foushee (Eds.), *Cockpit Resource Management Training* (pp. 15–22). Mountain View, CA: NASA Ames Research Center, National Aeronautics and Space Administration.
17. Shroyer, D. H. (1987). The development and implementation of CRM in UAL recurrent training. In H. W. Orlady and H. C. Foushee (Eds.), *Cockpit Resource Management Training* (pp. 47–50). Mountain View, CA: NASA Ames Research Center, National Aeronautics and Space Administration.
18. Hamman, W. R. (2010). Line oriented flight training (LOFT): The intersection of technical and human factor crew resource management (CRM) team skills. In B. G. Kanki, R. L. Helmreich and J. M. Anca (Eds.), *Crew Resource Management* (2nd edn., pp. 233–264). San Diego, CA: Academic Press.

19. Federal Aviation Administration. (2015). *Advisory Circular 120-35D: Flightcrew Member Line Operational Simulations: Line-Oriented Flight Training, Special Purpose Operational Training, Line Operational Evaluation*. Washington, DC: US Department of Transportation.
20. NTSB. (1988). Northwest Airlines Inc McDonnell Douglas DC-9-82 N312RC, Aviation Accident Report AAR-88-05. Washington, DC: National Transport Safety Board.
21. Federal Aviation Administration. (2006). *Advisory Circular 120-54A: Advanced Qualification Program*. Washington, DC: US Department of Transportation.
22. Weitzel, T. R., and Lehrer, H. R. (1992). A turning point in aviation training: The AQP mandates crew resource management and line operational simulations. *Journal of Aviation/Aerospace Education & Research*, *3*(1), 14–20.
23. Flin, R., Goeters, K., Hormann, H., and Martin, L. (1998). A generic structure of non-technical skills for training and assessment. Paper presented at the Proceedings of the 23rd Conference of the European Association for Aviation Psychology.
24. Flin, R., Martin, L., Goeters, K.-M., Hormann, H., Amalberti, R., Valot, C., and Nijhuis, H. (2003). Development of the NOTECHS (Non-Technical Skills) system for assessing pilots' CRM skills. *Human Factors and Aerospace Safety*, *3*, 97–120.
25. Flin, R., and Martin, L. (2001). Behavioural markers for crew resource management: A review of current practice. *International Journal of Aviation Psychology*, *11*(1), 95–118.
26. Howard, S. K., Gaba, D. M., Fish, K. J., Yang, G., and Sarnquist, F. H. (1992). Anesthesia crisis resource management training: Teaching anesthesiologists to handle critical incidents. *Aviation, Space, and Environmental Medicine*, *63*(9), 763–770.
27. Gaba, D. M., and DeAnda, A. (1988). A comprehensive anesthesia simulation environment: Re-creating the operating room for research and training. *Anesthesiology*, *69*(3), 387–394.
28. Chauvin, C., Lardjane, S., Morel, G., Clostermann, J.-P., and Langard, B. (2013). Human and organisational factors in maritime accidents: Analysis of collisions at sea using the HFACS. *Accident Analysis & Prevention*, *59*, 26–37.
29. Dupont, G. (1997). The dirty dozen errors in maintenance. Paper presented at the 11th FAA- AAM meeting on Human Factors in Aviation Maintenance and Inspection, San Diego, CA.
30. Flin, R. (1995). Crew resource management for teams in the offshore oil industry. *Journal of European Industrial Training*, *19*(9), 23–27.
31. Flin, R. (1996). *Sitting in the Hot Seat: Leaders and Teams for Critical Incident Management*. Chichester, UK: Wiley.
32. Shaw, A., Oakman, J., Thomas, M. J. W., Blewett, V., Aickin, C., Stiller, L., and Riley, D. (2014). *Associated Non Technical Skills: Action Learning Program Facilitators Resource*. Sydney, NSW: Department of Industry, Resources and Energy.

33. Nazir, S., Sorensen, L. J., Øvergård, K. I., and Manca, D. (2015). Impact of training methods on distributed situation awareness of industrial operators. *Safety Science*, *73*, 136–145.
34. Mavin, T. J., and Murray, P. S. (2010). The development of airline pilot skills through simulated practice. In S. Billett (Ed.), *Learning through Practice* (pp. 268–286). New York, NY: Springer.
35. Taylor, J. C. (2000). The evolution and effectiveness of maintenance resource management (MRM). *International Journal of Industrial Ergonomics*, *26*(2), 201–215.
36. Simpson, P., Owens, C., and Edkins, G. (2004). Cabin crew expected safety behaviours. *Human Factors and Aerospace Safety*, *4*(3), 153–167.
37. Andersen, V., and Bove, T. (2000). A feasibility study of the use of incidents and accidents reports to evaluate effects of team resource management in air traffic control. *Safety Science*, *35*(1), 87–94.
38. Krüger, A., Gillmann, B., Hardt, C., Döring, R., Beckers, S., and Rossaint, R. (2009). Teaching non-technical skills for critical incidents: Crisis resource management training for medical students. *Der Anaesthesist*, *58*(6), 582–588.
39. Flin, R., and Maran, N. (2004). Identifying and training non-technical skills for teams in acute medicine. *Quality and Safety in Health Care*, *13*(suppl 1), i80–i84.
40. Mitchell, L., Flin, R., Yule, S., Mitchell, J., Coutts, K., and Youngson, G. (2012). Evaluation of the scrub practitioners' list of intraoperative non-technical skills (SPLINTS) system. *International Journal of Nursing Studies*, *49*(2), 201–211.
41. Aggarwal, R., Undre, S., Moorthy, K., Vincent, C., and Darzi, A. (2004). The simulated operating theatre: Comprehensive training for surgical teams. *Quality and Safety in Health Care*, *13*(suppl 1), i27–i32.
42. Reader, T. W., Flin, R., Mearns, K., and Cuthbertson, B. H. (2009). Developing a team performance framework for the intensive care unit. *Critical Care Medicine*, *37*(5), 1787–1793.
43. Andersen, P. O., Jensen, M. K., Lippert, A., and Østergaard, D. (2010). Identifying non-technical skills and barriers for improvement of teamwork in cardiac arrest teams. *Resuscitation*, *81*(6), 695–702.
44. Shields, A., and Flin, R. (2013). Paramedics' non-technical skills: A literature review. *Emergency Medicine Journal*, *30*(5), 350–354.
45. Bahl, R., Murphy, D. J., and Strachan, B. (2010). Non-technical skills for obstetricians conducting forceps and vacuum deliveries: Qualitative analysis by interviews and video recordings. *European Journal of Obstetrics & Gynecology and Reproductive Biology*, *150*(2), 147–151.
46. Brewin, J., Tang, J., Dasgupta, P., Khan, M. S., Ahmed, K., Bello, F., Kneebone, R., and Jaye, P. (2015). Full immersion simulation: Validation of a distributed simulation environment for technical and non-technical skills training in urology. *BJU International*, *116*(1), 156–162.

47. Marshall, R. L., Smith, J. S., Gorman, P. J., Krummel, T. M., Haluck, R. S., and Cooney, R. N. (2001). Use of a human patient simulator in the development of resident trauma management skills. *Journal of Trauma and Acute Care Surgery*, *51*(1), 17–21.
48. Chalwin, R. P., and Flabouris, A. (2013). Utility and assessment of non-technical skills for rapid response systems and medical emergency teams. *Internal Medicine Journal*, *43*(9), 962–969.
49. O'Connor, P. (2011). Assessing the effectiveness of bridge resource management training. *International Journal of Aviation Psychology*, *21*(4), 357–374.
50. Darbra, R. M., Crawford, J., Haley, C., and Morrison, R. (2007). Safety culture and hazard risk perception of Australian and New Zealand maritime pilots. *Marine Policy*, *31*(6), 736–745.
51. Bonsall, K., and Pitsopoulos, J. (2012). The assessment and management of errors in a simulator. In J. R. Wilson, A. Mills, T. Clarke, J. Rajan and N. Dadashi (Eds.), *Rail Human Factors around the World* (pp. 656–665). Leiden, the Netherlands: CRC Press.
52. Lowe, A. R., Hayward, B. J., and Dalton, A. L. (2007). *Guidelines for Rail Resource Management*. Fortitude Valley, Australia: Rail Safety Regulators' Panel.
53. Flin, R., Wilkinson, J., and Agnew, C. (2014). *Crew Resource Management for Well Operations Teams*. London, UK: International Association of Oil & Gas Producers.
54. Flin, R., O'Connor, P., Mearns, K., Gordon, R., and Whittaker, S. (2002). Crew resource management training for offshore operations. Vol. 3 of 3. Factoring the human into safety: Translating research into practice (HSE Report OTO 2000 063). London, UK: HSE Books.
55. Martin, R., Maytham, B., Case, J., and Fraser, D. (2005). Engineering graduates' perceptions of how well they were prepared for work in industry. *European Journal of Engineering Education*, *30*(2), 167–180.
56. O'Connor, P., O'Dea, A., Flin, R., and Belton, S. (2008). Identifying the team skills required by nuclear power plant operations personnel. *International Journal of Industrial Ergonomics*, *38*(11), 1028–1037.
57. Crichton, M., and Flin, R. (2004). Identifying and training non-technical skills of nuclear emergency response teams. *Annals of Nuclear Energy*, *31*(12), 1317–1330.
58. Crichton, M. T., Flin, R., and Rattray, W. A. (2000). Training decision makers–tactical decision games. *Journal of Contingencies and Crisis Management*, *8*(4), 208–217.

3

Adult Learning Principles and Non-Technical Skills

So far, this book has explored the overarching need for non-technical skills training programs in high-risk industries and the historic context in which a number of these industries set about developing such training programs. We now turn to the practicalities of developing programs for training and assessing non-technical skills. At this stage, it is critical that we return to *first principles* and spend some time exploring the fundamental theories of training and assessment.

This chapter will explore the basic principles of adult learning theory and the range of factors we need to take into consideration when designing training programs for professionals operating in high-risk industries.

What Do We Mean by Training Non-Technical Skills?

As we have introduced already, there is a need for operators in high-risk industries to be able to draw on a range of knowledge, skills and attitudes that go beyond the technical knowledge and skills that enable specific tasks to be undertaken.

Take, for example, the control room operator who uses a supervisory control and data acquisition (SCADA) system to monitor a high-voltage electrical distribution network and change the status of pieces of equipment in the field to isolate or energise different aspects of the network. In being deemed competent to undertake this high-risk role, the control room operator will have undertaken a training program that involved developing a core body of knowledge about electricity and its transmission; he or she will have learnt about the operation of many types of equipment used across the distribution network, and will have learnt the technical aspects and standard operating procedures for operating the SCADA system in the control room. Many hours of study and guided practice would have been involved in developing the requisite knowledge, skills and attitudes for this job.

Unfortunately, it is likely that this control room operator has never received formal training in non-technical skills, but relies on their formal technical training and *informal* acquisition of skills in areas such as communication and decision-making to perform their tasks.

So far in this book, we have established that formal training in non-technical skills can result in enhanced performance. An operator who is trained in non-technical skills is positioned to communicate more effectively, make better decisions, and better manage the errors that will inevitably occur. However, what would this training involve, and what would it 'look like'? Would it be different from the types of training that developed their technical knowledge and skills? How would someone go about developing a non-technical skills training program for high-voltage electricity network control room operators? The next chapters will provide practical solutions to these problems.

On the Matter of Definitions

For the purposes of this book, we will be narrow in our use of terminology and adopt the term *training* to describe all activities associated with formal programs designed to develop non-technical skills in high-risk industries. We will also keep our heads below the parapet with respect to the debate around the various nuanced terms relating to learning, and will not attempt to differentiate between education and training, or any of the other terms used to describe activities that contribute to the development of new knowledge, skills and attitudes. Perhaps, as our domain of non-technical skills matures, we can begin to engage in those debates, but at this stage we will be more concerned with some basic first principles.

Of the different terms used in this book, *learning* is perhaps the most easily defined. Learning is the process of developing new knowledge, skills and attitudes. It is a ubiquitous human characteristic that continues throughout our lives. It is often associated with a framework of behavioural change, in terms of being able to do something we previously were unable to do.

For this book, *training* will be used to describe a collection of different types of learning activities, from the classroom seminar through to the use of high-fidelity simulation. Training is a process of guiding trainees through purposeful activity to develop new knowledge, skills and attitudes. Training is typically tightly focussed on clearly defined learning objectives that have been created to meet specific aspects of performing a job.[1]

We will also be narrow in our use of terms to describe the participants engaged in the training; we will use the term *instructor* to describe those who assist in the process of learning, and we will use *trainee* to describe those who are developing new knowledge, skills and attitudes.

Principles of Adult Learning

The design of non-technical skills training programs in high-risk industries must adopt an approach that understands the specific needs of trainees. Unlike traditional formal education programs, non-technical skills training programs are usually developed for adult learners who bring a significant pre-existing set of skills in many of the domains of non-technical skills. Accordingly, non-technical skills training programs need to recognise this context and turn to adult learning theory to develop the most effective training interventions.

During the 1970s, educational psychologists such as Malcolm Knowles developed a more formalised approach to the ways in which adults differ from children in their approach to learning. While considerable research had guided modern educational theory and practice, most of this research related to the formal education of children in schools. However, adults have specialised needs and bring to training and education a wealth of pre-existing knowledge and skill.

Adult learning theory sets out six main principles to guide the development of training and education for adults who bring to the learning experience significant pre-existing knowledge and skill.[2] These principles are summarised in Table 3.1.

Individual Learning Styles

Another important consideration in adult learning is that not all individuals learn in the same way, and that everyone has their own preferences in terms of learning. These individual differences are called

Table 3.1 Principles of Adult Learning Theory

PRINCIPLE	DESCRIPTION
Adult learners need to have learning contextualised	Adults like to have the reason for learning explained to them prior to investing the requisite energy into the learning process.
Adult learners are autonomous and self-directed	Adults don't really like being told what to do, and are more open to guidance and facilitation, such that they can self-direct their own learning.
Adult learners bring with them considerable experience	Adults have considerable experience, on which new knowledge and skills are built.
Adult learners need a reason to learn	Adults are ready to learn when they perceive a need for developing new knowledge and skills.
Adult learners are real-world problem centred	Adult learners focus on learning new knowledge and skills that will enable enhanced job performance in the real world.
Adult learners have intrinsic motivation	Rather than relying on external forces to motivate learning, adults are more inclined to be internally motivated.

Source: Knowles, M.S., *The Adult Learner: A Neglected Species*, Houston, TX, Gulf Publishing Company, 1990.

learning styles, and describe the different ways in which people process information in the training context.[3]

Although myriad different theories of individual learning styles can be found in the education and training literature, many share common elements. One of the most popular theories of learning style was developed by David Kolb, who sets out four types of preferred learning style based on the dimensions of *perception* (feeling and experiencing vs. thinking and conceptualising) and *processing* (watching and observing vs. actively doing).[4] Table 3.2 provides a summary of these four

Table 3.2 Kolb's Four Learning Styles and Aligned Strategies for Training Design

LEARNING STYLE	ALIGNED TEACHING/LEARNING STRATEGIES
Divergers (Type 1) Feeling: Experience Watching: Observation	Listening Watching role-play and simulation Group discussion
Assimilators (Type 2) Thinking: Conceptualising Watching: Observation	Seminar Case studies eLearning
Convergers (Type 3) Thinking: Conceptualising Doing: Active experimentation	Problem-based learning Putting theory into practice
Accommodators (Type 4) Feeling: Experience Doing: Active experimentation	Participating in role-play/participating in simulation

types of learning, and some training strategies that are appropriate in the context of non-technical skill development.

One important consideration is that these learning styles are *preferred* styles only, and an individual can adapt to learning in non-preferred styles. One consideration of training system design is that a simple learning styles inventory can be used to identify individual preferences of learners and shape subsequent instruction to maximise learning aligned with an individual's preferred style. Similarly, offering a range of different learning activities can increase the overall effectiveness of a training program such that individuals can experience learning both within and outside their preferred style.

Domains of Learning

Learning is by no means a mono-dimensional construct, but rather, a set of processes that differ as a function of the type of learning outcomes being achieved. An effective training program specifies the type of learning outcome being developed and uses an appropriate training strategy that matches the variety of learning. For instance, it is simply not possible to develop the fine motor skills used for landing an aircraft or performing a surgical procedure through a training approach that simply builds new knowledge. Rather, motor skills are developed by repeated practice in performing the actual task.

There are a number of ways in which the domains of learning have been categorised. One of the first of these was put forward by Bloom and his colleagues in the 1950s.[5] Bloom's taxonomy described the cognitive domain (knowledge), the affective domain (attitudes) and the psychomotor domain (skills). Each of these domains is then further broken down, and appropriate learning strategies are suggested. This *knowledge – skills – attitudes* (KSA) differentiation is still very common today. This specification of the domains of learning and the basic instructional strategies that support them is provided in Table 3.3.

With respect to non-technical skills, we are attempting to develop a range of knowledge, cognitive skills and attitudes. Accordingly, our instructional design process needs to take into consideration the type of learning outcome we are trying to achieve, and use an appropriate instructional strategy to do so. The most important consideration to reinforce at this point is that no non-technical skills training program

Table 3.3 Domains of Learning Outcomes

DOMAIN	DESCRIPTION AND TRAINING STRATEGY
Cognitive outcomes	*Knowledge:* An understanding of concepts and facts. *Cognitive strategies:* Ways of processing information. *Cognitive outcomes are developed through associating new ideas with pre-existing knowledge, and through working on new problems that present challenges to thinking.*
Skill-based outcomes	*Skills:* Being able to perform a task accurately and consistently. *Automaticity:* Being able to perform a task without constantly thinking about actions. *Skill-based outcomes are developed through repeated practice of a particular skill in different situations.*
Affective outcomes	*Attitudes:* A set of beliefs that influences choices about actions. *Motivation:* An influence that promotes specific actions. *Affective outcomes are developed and modified indirectly through knowledge acquisition, behavioural modelling and reinforcement.*

Sources: Gagné, R. et al., *Principles of Instructional Design*, New York, NY, Harcourt Brace Jovanovich College Publishers, 1992 and Kraiger, K. et al., *J. Appl. Psychol.*, *78*(2), 311, 1993.

can rely on classroom-based training to achieve learning outcomes across all domains of learning.

The Enabling Role of Core Knowledge

For any non-technical skills training program to be effective, it must first ensure that participants have the requisite knowledge to understand the nature of the skills being developed.[8] In traditional training, there has always been a clear relationship between theoretical knowledge and skill development. Indeed, the basic understanding of a skilled practitioner is often referred to as someone who is good at putting theory into practice. In many professional domains, theoretical knowledge is taught prior to practical skill development.

CASE STUDY: PUTTING THEORY INTO PRACTICE FOR LAPAROSCOPIC SURGERY

Laparoscopic surgery, sometimes known as 'keyhole' surgery, involves the use of a camera and a surgical instrument being inserted through only very small incisions. The surgeon guides with one hand a camera on the end of a rod, and the other hand guides the required surgical instrument on the end of a rod.

The development of skills in laparoscopic surgery is a classic example of the enabling role of theoretical knowledge in the context of psychomotor skill development.

Prior to learning the art of laparoscopic surgery, the surgeon must have learnt the requisite knowledge with respect to anatomy, the relevant disease, as well as a host of other pieces of theoretical medical knowledge. Further, the surgeon must also have knowledge relating to the technical operation of the laparoscope and associated surgical equipment. This knowledge is then *integrated* with technical psychomotor skills in operating the equipment to actually undertake a safe and successful operation.

While this example focusses on technical knowledge and skill, exactly the same approach can be taken with respect to the training of non-technical skills.

The Reflective Practitioner

Another way of describing an appropriate orientation for the high-risk industry worker enhancing their non-technical skills is that of the *reflective practitioner*. This term was coined by Donald Schön to describe the approach adopted by professionals to develop enhanced skills across their careers.[9]

Reflective practice was developed in contrast to the traditional training approach termed *technical rationality*, which suggests that professional practice involves the application of technical knowledge and concrete skills that have been acquired through formal training programs.[10]

In many respects, reflective practice aligns nicely with the concept of non-technical skill development. The learning approaches described by technical rationality reflect how, until very recently, technical knowledge and skill were given the highest priority. The concept of reflective practice shows that there is much more to achieving safe and efficient performance in high-risk industries than technical skill alone.

A body of research has picked up on Shön's concept of the reflective practitioner and has described how it can inform skill development in the industrial training context and also in high-risk industries. However, experience in industries such as healthcare highlights that reflective practice does not simply occur naturally, but rather, requires some formal structures to occur in the everyday work environment.[11] As will be explored in detail in this and upcoming chapters, such support takes the form of guided experiences, in which reflective practice is promoted, and through facilitated debriefing of those experiences.

The basic premise of the reflective practitioner concept as it relates to the development of non-technical skills is that training programs must provide authentic real-world scenarios in which professional practice is performed, and then that practice must be critically reflected on to enhance future performance. Furthermore, another important aspect of training programs that support reflective practice is that they allow trainees to test their knowledge, skills and attitudes in front of their peers. This type of learning (termed *Model II learning*) is different from the more private forms of self-study promoted in many industrial contexts, as it promotes the social-normalisation of professional practice and can address issues of inappropriate attitudes or assumptions.[12]

Since Shön's original model, subsequent developments have been made to reconcile the approaches to technical skill development with those of reflective practice to build a more comprehensive model of on-going professional development.[13]

Constructivism

Perhaps one of the most relevant learning theories for the training and assessing of non-technical skills is *constructivism*. This theory of learning suggests that learning is the active process of an individual constructing meaning. Constructivism as a learning theory owes much to the work of psychologists Jean Piaget[14] and Lev S. Vygotsky.[15]

Constructivism highlights that we learn best when we first build a scaffold around which further knowledge and skill development can be built. Further, constructivism highlights that learning is not only the process of an individual developing more and more knowledge and skill; also, this development takes place in a social context. To this end, constructivism emphasises a social negotiation of knowledge, skills and attitudes, and is also a process of enculturation.[16]

This approach to learning is another that presents a critique of technical rationality, and it is aligned with the understanding that human intelligence, with its intuitive and inductive characteristics and its social origins, goes far beyond simply rule-based skills.[17] This is an important consideration with respect to the development of non-technical skills in high-risk industries, where a specific *orientation* to performance and safety must be maintained. This orientation is wrapped up in individual attitudes as well as organisational, professional and national cultures.

Much has been made of a set of appropriate attitudes and the overarching safety culture of an organisation, and a significant body of research now underpins these assertions. However, much less has been made of how these attitudes and values are created and sustained within an organisation. The social constructivist approach to adult learning can offer much in this regard, and to this end, the non-technical skills training program is a potentially powerful vehicle for attitudinal and cultural change to enhance safety. It is beyond the scope of this book to explore this concept in further detail. However, suffice it to say that without necessarily promoting a return to the explicit 'indoctrination' programs of early crew resource management training, non-technical skills training programs do play a role in attitudinal and cultural change. Accordingly, the underlying attitudes and values of any high-risk organisation should be made explicit as part of the non-technical skills training curriculum.

Stages of Instruction

One of the potential dangers associated with learning complex skills, such as those associated with non-technical domains, is that learning can actually be inhibited due to trainees becoming overwhelmed by task complexity and cognitive load. To overcome this potential problem, several strategies have been suggested. Of these, the strategy of simple to complex task sequencing is of considerable benefit to non-technical skills training programs. It builds on the concept of scaffolding; task demands are kept manageable by working on simplified part-tasks before integrating those skills into more complex authentic scenarios.[18]

Linked closely to the concepts embodied in constructivism, modern models of instruction emphasise a staged process of knowledge and skill development. The most common of these breaks the process of instruction down into four distinct phases: (1) the activation of a trainee's prior knowledge, skills and experience; (2) demonstration of new skills; (3) the trainee's application of those skills; and (4) integration of those skills into real-world activities.

Anyone familiar with the process of technical skill development in flight training, or in the training of surgical techniques, will instantly recognise this as the dominant approach in current training programs. The use of simulation in training, and the common progression from lower-fidelity part-task trainers to higher-fidelity 'full mission' simulators, and then into the real world, also reflects this concept of a staged approach to instruction.[19]

The approach to non-technical skills development should be no different, and in the following chapter, we will examine how we can map out staged skill development from the underlying basic cognitive or social skills into more complex integrated non-technical performance.

Elaboration Theory

Another approach to sequencing instruction is that of elaboration theory, which adopts a model of successively transitioning from the general to the detailed.[20] Elaboration theory adopts the analogy of someone using a zoom lens. First, a wide angle is used to get the big picture and overall context of the scene, before zooming in to specific areas to gather a more detailed view of individual parts. The proponents of elaboration theory suggest that instruction has typically adopted a zoom lens from the outset, progressively moving from concept to concept at a high level of detail. This has been shown to prevent effective integration of new knowledge, skills and attitudes, and also has negative implications for transfer of training, more of which we will explore shortly.

The basic principle of elaboration theory is that new knowledge, skills and attitudes should be presented in the broader context prior to being elaborated on in more and more detail, and then that learning should be synthesised back into the broader context. In the context of professional training, the instructional approach of elaboration is critical to manage the cognitive load of the trainee and to ensure that the process of learning is not hindered by overload, as will be discussed shortly.[21]

Cognitive Load Theory

Another important consideration relating to adult learning theory is the management of cognitive load throughout the training program. It has been demonstrated consistently that learning is inhibited when trainees are thrown in 'at the deep end' and asked to demonstrate complex skills without a progressive and staged development of those skills and the knowledge that underpins them.[22]

One approach to managing cognitive load that is relevant to non-technical skills training programs is that of beginning with a 'worked-out example', in which the actions of individuals and teams that contributed to a positive or negative outcome, and the underlying knowledge, skills and attitudes that gave rise to those actions, are made explicit to trainees.[23] In this way, the initial cognitive load is minimised by the 'working out' and the 'correct answers' being provided to trainees in the early stages of instruction.

Such an approach can easily be integrated into case-based modes of training as an important element of the sequence from basic knowledge and skill acquisition through to practice and competent performance.

Modelling Behaviour

The process of observing and reproducing behaviours through imitation is a form of learning that is critical to social and cultural development, even from the very early stages of childhood development.[24] Often termed *observational learning*, *social learning* or *social modelling*, it has been demonstrated that individual safety-related behaviour is highly influenced by what a trainee sees as accepted group norms.[25]

Many of the adult learning principles described in the preceding sections highlight the role of *action* in learning non-technical skills. Indeed, we learn much from our experience of the negative and positive outcomes associated with specific actions. However, we as humans have an advanced capacity to develop new knowledge and skills simply by observing others. This capacity is not simply mimicry, but rather, involves reflection on the action of others and the subsequent generation of abstract rules governing specific judgements and behaviours.[26]

Situated Learning and Cognitive Apprenticeships

The term *cognitive apprenticeship* was coined by Allan Collins, John Seely Brown and Susan Newman in the late 1980s to describe a novel approach to teaching basic skills in reading, writing and mathematics.[27] The approach moved away from the traditional didactic model to one in which cognitive skills were developed through working on real-world tasks and through a staged process of observation, coaching and practice.

Cognitive apprenticeship models have more recently been emphasised in vocational training and as an important approach to skill development in high-risk industries such as medicine.[28] Picking up on the theme of enculturation, which is made explicit in theories such as constructivism, the notion of a cognitive apprenticeship suggests that trainees are inducted into a community of practice.[29] Often, these aspects have been part of a 'hidden' curriculum,[30] but the domain of non-technical skills requires us to make these aspects explicit and ensure that they are embedded within our overall instructional design.

One of the critical aspects of cognitive apprenticeships is making explicit the tacit cognitive activities undertaken by experts. This approach has much to offer to the on-the-job modes of non-technical skill training, and emphasises that trainees learn not just by observing and doing, but also by learning the underlying cognitive strategies used by experts to optimise performance in high-risk domains.

Closely linked to the concept of cognitive apprenticeships is the concept of situated learning. The theory of situated learning emphasises that learning must take place in authentic contexts. This is particularly relevant to non-technical skill development, as situated learning holds that the appropriateness of any action can only be described in the context in which it is used, and that skill development must be situated within authentic contexts. Further, these contexts must be seen not only in task-specific terms but also in the complex and social environment of day-to-day operations.[31]

Instructional Use of Error: Learning from Our Mistakes

While the idiom of 'trial and error' describes informal learning processes, exposure to error during formal training has been identified as highly beneficial to subsequent skill development. Making errors during the process of skill development forces greater conscious reflection on performance and engagement with learning.[32] It enables learners to see the negative outcomes of their actions and therefore prompts the identification of corrective strategies.[33] This training approach is especially effective if avoidance or corrective strategies are debriefed by a domain expert, who is able to guide the trainees to aspects of optimal performance.

The use of error as an instructional tool has been highlighted as an important element of training high-reliability teams and with specific reference to non-technical competencies.[34] Experiential forms of learning in which it is possible to make errors without compromising safety, such that lessons can be drawn to avoid similar errors in the future, are important aspects of training. Further, responding to error effectively is an essential aspect of safe performance in high-risk industries, and the skills that enable us to do so are non-technical in nature.[35,36]

For instance, in simulation-based medical education, the basic assumption is that increased practice in learning from error and in the management of everyday error in a simulated environment will reduce occurrences of errors in real life and will provide professionals with the correct attitude and skills to cope competently with those errors when they do unavoidably occur.[37] This approach to training is quite different from more traditional approaches of competency development.

The use of error in learning is a well-known informal learning strategy that comes naturally to us as early as our infant years. Throughout our lives, we make errors, reflect on how and why we made those errors, and work out how to change our actions in the future. Therefore, learning in an environment where it is acceptable and safe to make errors is important when training and assessing non-technical skills.

One of the unique characteristics of learning is that we can also learn from other people's mistakes. As will be discussed in the following chapter, the use of case studies where a negative outcome occurs enables trainees to examine the underlying reasons *why* sub-optimal performance resulted in a negative outcome.

The use of simulation in non-technical skills training provides a unique environment in which error can be encountered, or even promoted, without an adverse impact on safety. In the simulator, not only is the actual physical safety of the trainees and instructors never compromised, but the environment is one where psychological safety can also be maintained. In many domains, such as healthcare, this notion of being able to safely make errors in the learning process is critical. Traditionally, many skills were developed by trainee doctors through interacting with real patients. While this medical apprentice model is long-standing and has many benefits, the reality was that any error by a trainee doctor could result in actual harm to the patient. In this context, medical simulation in its many forms provides a unique environment where error can be harnessed safely as a mechanism for enhanced learning.[38]

By emphasising the role of error in learning and legitimising the fact that error is likely to occur, enhanced training outcomes can be realised. The use of simulation in non-technical skills training programs will be explored in more detail in the following chapters.

Competency-Based Training

Competency-based training, or, as referred to in some domains, competency-based education, has become deeply entrenched as the most common approach to vocational skill development in high-risk industries.

Competency-based training describes an approach that focuses specifically on the development of specific skills that have been empirically established as critical to expert task performance. Training is structured around strategies for developing those skills, and assessment focuses on determining that a criterion level of performance can be consistently demonstrated. A recent systematic review of the defining features of competency-based training sets out a series of common elements of competency-based training programs.[39] Some of these key features are summarised and elaborated on in Table 3.4.

In competency-based training, there is a need to define the point at which competency is achieved, based on observable performance. The time it takes to achieve competency may vary considerably from individual to individual. We need to take this factor into consideration when designing training programs for non-technical skills, and not be tied to single training events in our training program design.

Transfer of Training

The *transfer of training* refers to the degree to which skills learnt in one training setting, such as the classroom or simulator, can be transferred to other different tasks or contexts and into the 'real world' of day-to-day operations. This concept has been the focus of educational and psychological research for over a century, and has been the subject

Table 3.4 Summary of the Key Features of Competency-Based Training

STAGE OF DECISION-MAKING	DESCRIPTION
Taxonomy of competencies	Competency-based training is centred around a pre-determined set of competencies that together describe expert performance. These competencies are derived from a range of sources, including detailed task and training needs analyses, observation of, and interviews with, domain experts, and established standards. In the training domain, we call these taxonomies of competencies a *competency specification.*
Criterion-referenced assessment	Competency-based training uses formative and summative criterion-referenced assessment to assess progress and certify that proficiency in the skill is demonstrated. Criterion-referenced assessment involves evaluating performance of a trainee against pre-determined criteria of what actions and outcomes are considered as demonstrating competence.
Learner-centred	Competency-based training is by its very nature learner-centred. It adopts approaches to curriculum design that focus on trainee skill development, rather than the didactic presentation of information.
De-emphasises time spent training	Competency-based training purposefully de-emphasises the specification of time spent developing a skill; rather, it accepts that different trainees will take different amounts of time to demonstrate competence. For instance, a surgical trainee might need to practice for two hours on a laparoscopic skills development simulator to achieve the criterion level of performance, whereas another colleague may achieve that level in one hour.

of research by such eminent researchers as Edward Thorndike and Robert M Gagné.

Of importance to our study of training and assessing non-technical skills, recent reviews of empirical research have reinforced that student motivation and training context is critical to the transfer of non-technical skills training programs.[40] Also, the concepts of training scenario design and simulation fidelity are critical to the effective transfer of training, and these will be explored in the following chapters.

Adult Learning: A Summary

This section has briefly introduced the primary theories of adult learning that are relevant to the design of non-technical skills training programs. They represent the core concepts that need to be applied in the design of effective training programs for adult learners operating in high-risk industries. Table 3.5 provides a brief checklist to refer to when designing non-technical skills training programs to ensure that these core concepts are put into practice.

Table 3.5 Checklist for Putting Adult Learning Theory into Practice

CHECKLIST ITEM	✓
PRINCIPLES OF ADULT LEARNING	
The reason for learning is explained to trainees prior to them investing energy into the learning process.	☐
Training adopts an approach of guidance and facilitation, such that trainees can self-direct their own learning.	☐
There has been an analysis of the experience trainees bring to the training program, on which new knowledge and skills are built.	☐
Trainees recognise a need for developing new knowledge and skills.	☐
The training program focusses on new knowledge and skills that will enable enhanced job performance in the real world.	☐
The training program is designed in recognition of the fact that trainees are more inclined to be internally motivated.	☐
INDIVIDUAL LEARNING STYLES	
The training program takes into consideration the preferred learning style of trainees.	☐
DOMAINS OF LEARNING	
Each element of the training program is appropriate to the domain of learning outcome. For instance, knowledge is developed through reading, seminars and workshops, skills are developed through simulation and on-the-job practice, and attitudes are developed through critical reflection and facilitated discussion.	☐
REFLECTIVE PRACTICE	
The training program promotes non-technical skill development through processes of reflective practice alongside traditional skills-based training approaches such as part-task rehearsal.	☐
The training program uses authentic scenarios to create realistic environments for professional practice.	☐
The training program promotes critical reflection by trainees on their own performance and that of the team.	☐
CONSTRUCTIVISM	
The training program promotes non-technical skill development as a process of learning in a social and cultural context.	☐
The training program makes explicit the underlying assumptions of attitudes and cultural values that contribute to safe and efficient operations.	☐
STAGES OF INSTRUCTION	
The training program is designed in a manner that ensures learning is facilitated through appropriate levels of task complexity and cognitive load.	☐
Each element in the training program adopts a sequence of instruction that includes: (1) the activation of a trainee's prior knowledge, skills and experience; (2) demonstration of new skills; (3) the trainee's application of those skills; and (4) integration of those skills into real-world activities.	☐

(*Continued*)

Table 3.5 (Continued) Checklist for Putting Adult Learning Theory into Practice

CHECKLIST ITEM	✓
ELABORATION THEORY	
The training program first contextualises new knowledge, skills and attitudes prior to elaborating on them in increasingly fine levels of detail.	☐
Once the knowledge, skills and attitudes have been developed in detail, this learning is then generalised across the larger context of day-to-day work.	☐
COGNITIVE LOAD THEORY	
The training program is designed to manage cognitive load; it does not ask trainees to demonstrate complex skills without first developing the underlying enabling knowledge and constituent skills.	☐
Strategies such as using 'worked examples' in case-based training modes are integrated into the sequence of training.	☐
MODELLING BEHAVIOUR	
The training program provides exposure to examples of good and poor non-technical performance, such that behaviours can be observed and modelled.	☐
SITUATED LEARNING	
The training program is designed to enable learning to take place in authentic contexts, including the social and cultural contexts of the real-world work environment.	☐
The training program makes explicit the tacit cognitive strategies used by experts in managing ill-defined real-world problems.	☐
INSTRUCTIONAL USE OF ERROR	
The training program adopts an approach in which error is expected to occur and is used to enhance learning by highlighting sub-optimal strategies.	☐
COMPETENCY-BASED TRAINING	
The training program specifies a taxonomy of competencies to be developed through the training, and these competencies are developed prior to designing the curriculum.	☐
The training program adopts criterion-referenced assessment.	☐
The training program is learner-centred and focuses on individual needs and competency development.	☐
The training program allows flexibility in the time spent on developing competencies, and training does not finish until competency is attained.	☐

References

1. MacLeod, N. (2001). *Training Design in Aviation*. Aldershot, UK: Ashgate Publishing.
2. Knowles, M. S. (1990). *The Adult Learner: A Neglected Species* (4th edn.). Houston, TX: Gulf Publishing Company.

3. Henely, I. M. A., and Bye, J. (2003). Learning styles, multiple intelligences and personality types. In I. M. A. Henely (Ed.), *Aviation Education and Training: Adult Learning Principles and Teaching Strategies* (pp. 92–117). Aldershot, UK: Ashgate Publishing.
4. Kolb, D. (1976). *Learning Styles Inventory*. Boston, MA: McBer.
5. Bloom, B. S. (1956). *Taxonomy of Educational Objectives: The Classification of Educational Goals*. New York, NY: Longmans Green.
6. Gagné, R., Briggs, L. J., and Wager, W. W. (1992). *Principles of Instructional Design*. New York, NY: Harcourt Brace Jovanovich College Publishers.
7. Kraiger, K., Ford, J. K., and Salas, E. (1993). Application of cognitive, skill-based, and affective theories of learning outcomes to new methods of training evaluation. *Journal of Applied Psychology*, *78*(2), 311.
8. Civil Aviation Authority. (2014). *Flight-Crew Human Factors Handbook: CAP 737*. London, UK: Civil Aviation Authority UK.
9. Schön, D. (1983). *The Reflective Practitioner*. New York, NY: Basic Books.
10. Schön, D. (2002). From technical rationality to reflection-in-action. In R. Harrison, F. Reeve, A. Hanson and J. Clarke (Eds.), *Supporting Lifelong Learning: Perspectives on Learning* (pp. 40–61). New York, NY: Routledge Falmer.
11. Powell, J. H. (1989). The reflective practitioner in nursing. *Journal of Advanced Nursing*, *14*(10), 824–832.
12. Schön, D. (1987). *Educating the Reflective Practitioner*. San Francisco, CA: Jossey-Bass.
13. Cheetham, G., and Chivers, G. (1998). The reflective (and competent) practitioner: A model of professional competence which seeks to harmonise the reflective practitioner and competence-based approaches. *Journal of European Industrial Training*, *22*(7), 267–276.
14. Piaget, J. (1973). *Memory and Intelligence*. New York, NY: Basic Books.
15. Vygotsky, L. S. (1978). *Mind in Society: The Development of Higher Psychological Processes*. Cambridge, MA: Harvard University Press.
16. Merriam, S. B., Caffarella, R. S., and Baumgartner, L. M. (2007). *Learning in Adulthood: A Comprehensive Guide* (3rd edn.). San Francisco, CA: Jossey-Bass.
17. Dreyfus, H., and Dreyfus, S. (1986). *Mind over Machine: The Power of Human Intuition and Expertise in the Era of the Computer*. New York, NY: Free Press.
18. Van Merriënboer, J. J., Kirschner, P. A., and Kester, L. (2003). Taking the load off a learner's mind: Instructional design for complex learning. *Educational Psychologist*, *38*(1), 5–13.
19. Merrill, M. D. (2002). First principles of instruction. *Educational Technology Research and Development*, *50*(3), 43–59.
20. Reigeluth, C. M., Merrill, M. D., Wilson, B. G., and Spiller, R. T. (1980). The elaboration theory of instruction: A model for sequencing and synthesizing instruction. *Instructional Science*, *9*(3), 195–219.

21. Van Merriënboer, J. J., and Sweller, J. (2010). Cognitive load theory in health professional education: Design principles and strategies. *Medical Education*, *44*(1), 85–93.
22. Sweller, J. (1994). Cognitive load theory, learning difficulty, and instructional design. *Learning and Instruction*, *4*(4), 295–312.
23. Renkl, A., and Atkinson, R. K. (2003). Structuring the transition from example study to problem solving in cognitive skill acquisition: A cognitive load perspective. *Educational Psychologist*, *38*(1), 15–22.
24. Gergely, G., and Csibra, G. (2005). The social construction of the cultural mind: Imitative learning as a mechanism of human pedagogy. *Interaction Studies*, *6*(3), 463–481.
25. Olson, R., Grosshuesch, A., Schmidt, S., Gray, M., and Wipfli, B. (2009). Observational learning and workplace safety: The effects of viewing the collective behavior of multiple social models on the use of personal protective equipment. *Journal of Safety Research*, *40*(5), 383–387.
26. Bandura, A. (1986). *Social Foundations of Thought and Action: A Social Cognitive Theory*. Englewood Cliffs, NJ: Prentice-Hall.
27. Collins, A., Brown, J. S., and Newman, S. E. (1988). Cognitive apprenticeship. *Thinking: The Journal of Philosophy for Children*, *8*(1), 2–10.
28. Bleakley, A. (2006). Broadening conceptions of learning in medical education: The message from teamworking. *Medical Education*, *40*(2), 150–157.
29. Lave, J., and Wenger, E. (1991). *Situated Learning: Legitimate Peripheral Participation*. Cambridge, UK: Cambridge University Press.
30. McNeil, H. P., Hughes, C. S., Toohey, S. M., and Dowton, S. B. (2006). An innovative outcomes-based medical education program built on adult learning principles. *Medical Teacher*, *28*(6), 527–534.
31. Anderson, J. R., Reder, L. M., and Simon, H. A. (1996). Situated learning and education. *Educational Researcher*, *25*(4), 5–11.
32. Ivancic IV, K., and Hesketh, B. (2000). Learning from errors in a driving simulation: Effects on driving skill and self-confidence. *Ergonomics*, 43(12), 1966–1984.
33. Lorenzet, S. J., Salas, E., and Tannenbaum, S. I. (2005). Benefiting from mistakes: The impact of guided errors on learning, performance, and self-efficacy. *Human Resource Development Quarterly*, *16*(3), 301–322.
34. Wilson, K. A., Burke, C. S., Priest, H. A., and Salas, E. (2005). Promoting health care safety through training high reliability teams. *Quality and Safety in Health Care*, *14*(4), 303–309.
35. Thomas, M. J. W. (2004). Predictors of threat and error management: Identification of core non-technical skills and implications for training systems design. *International Journal of Aviation Psychology*, *14*(2), 207–231.
36. Thomas, M. J. W., and Petrilli, R. M. (2012). Error detection during normal flight operations: Resilient systems in practice. In A. de Voogt and T. D'Oliveira (Eds.), *Mechanisms in the Chain of Safety* (pp. 107–115). Aldershot, UK: Ashgate Publishing.

37. Ziv, A., Ben-David, S., and Ziv, M. (2005). Simulation based medical education: An opportunity to learn from errors. *Medical Teacher*, *27*(3), 193–199.
38. Ziv, A., Wolpe, P. R., Small, S. D., and Glick, S. (2006). Simulation-based medical education: An ethical imperative. *Simulation in Healthcare*, *1*(4), 252–256.
39. Frank, J. R., Mungroo, R., Ahmad, Y., Wang, M., De Rossi, S., and Horsley, T. (2010). Toward a definition of competency-based education in medicine: A systematic review of published definitions. *Medical Teacher*, *32*(8), 631–637.
40. Blume, B. D., Ford, J. K., Baldwin, T. T., and Huang, J. L. (2010). Transfer of training: A meta-analytic review. *Journal of Management*, *36*(4), 1065–1105.

4

Principles of Training Non-Technical Skills

Introduction

Training programs designed specifically to develop non-technical skills can take many forms, and there is certainly no single 'best practice' for such programs. However, as we have seen in the opening chapters of this book, research examining the effectiveness of non-technical skills training programs only began to be undertaken within the last few decades. Moreover, the majority of such training programs have drawn heavily on aviation crew resource management (CRM) for their inspiration and design. Accordingly, there are only a few validated approaches to training design from which all high-risk industries can draw. However, non-technical skills training programs are constantly evolving, and the next few decades will see significant advancement in training techniques.

In this chapter, we will build on the adult learning theory introduced in the previous chapter and introduce a range of practical methods for training non-technical skills. The important role of underpinning knowledge as a foundation on which skill development can occur is a common theme in non-technical skills training programs.[1] On this foundation of core knowledge, skill development can then take place.

This chapter will introduce a set of principles that can be used in the design of non-technical skills training programs. In the context of each of these principles, practical advice is provided for maximising the effective use of different training approaches prior to exploring assessment methodologies and how these can all be integrated into an overall curriculum for non-technical skills training in the following chapters. The principles introduced in this chapter are as follows:

- Principle One: Training of non-technical skills involves a phased approach.
- Principle Two: There are benefits to integrating non-technical skills into existing training programs.
- Principle Three: Non-technical skills training programs should consider the multidisciplinary and team-based nature of work in their design.
- Principle Four: Consideration should be given to the scope of non-technical skills training in terms of being generic or task specific.
- Principle Five: Attitudinal change is an important outcome of non-technical skills training programs.
- Principle Six: Core knowledge about non-technical skills needs to be built as a foundation for skill development.
- Principle Seven: There are significant limitations to the use of classroom and eLearning modes in non-technical skills training programs.
- Principle Eight: Simulation presents a highly effective training mode for the development of non-technical skills.
- Principle Nine: Non-technical skills training need not be restricted only to formal training programs.
- Principle Ten: Briefing and debriefing are essential components of any non-technical skills training session.

A Phased Approach to Training of Non-Technical Skills

The basic approach to the training of non-technical skills involves three primary phases. First, *awareness* of the importance of non-technical skills needs to be developed. The process of raising awareness of the critical nature of non-technical skills is fundamental to activating the necessary motivational focus for knowledge and skill development that we explored in the previous chapter. Trainees need to see the value of engaging in non-technical training programs and see the relevance of training to their own work.

The second phase of a non-technical skills training program involves *core enabling knowledge* being developed. This core enabling knowledge provides the theoretical underpinnings for skill development and introduces the conceptual framework on which skills can be built.

The final phase of a non-technical skills training program focuses on detailed skill development. First, a set of *basic skills* is developed, each of which forms part of a domain of non-technical skill such as communication. Finally, those basic skills are integrated into *actual task performance*.

As we have introduced in previous chapters, the traditional classroom-based training programs, such as the CRM seminar, are insufficient for skill development and behavioural change. Rather, a combination of classroom sessions, opportunities to practice behaviours in authentic settings, and expert feedback and reflection on those practice sessions are the main ingredients for a successful non-technical skills training program.[2] Having escaped the bounds of the classroom, most non-technical skills training programs are structured around a curriculum for both knowledge development and skill development. In domains such as healthcare, this model has been significantly refined in some specialties, and a *blended* approach has been created whereby training participants move backwards and forwards between the traditional seminar format and more experiential learning environments such as the simulator.[3]

PRINCIPLE ONE: TRAINING OF NON-TECHNICAL SKILLS INVOLVES A PHASED APPROACH

A non-technical skills training program focuses on three key phases, starting with raising awareness of the important role of non-technical skills, through the development of core enabling knowledge, and finally developing the non-technical skills themselves.

Training Context of Non-Technical Skills Programs

Every high-risk industry across the globe already has a set of clearly specified training programs that operators must complete on a regular basis. Therefore, each new industry that embraces non-technical skills must situate the new forms of training in an already well-established training environment. As many new non-technical skills training programs borrow much from the original forms of aviation CRM programs, this has often meant that non-technical skills training programs are introduced as stand-alone training programs.

This approach can be seen to constrain the effectiveness of a non-technical skills training program from a number of perspectives. First, adding an additional training program into an environment where training budget and training time are already constrained can limit the investment into the new non-technical skills training program. Second, a stand-alone program is limited in the degree to which non-technical skills and technical skills can be trained in an integrated manner. Accordingly, the ability to integrate a non-technical skills training program into existing forms of training can provide significant benefits to the organisation.

PRINCIPLE TWO: THERE ARE BENEFITS TO INTEGRATING NON-TECHNICAL SKILLS INTO EXISTING TRAINING PROGRAMS

In designing a non-technical skills training program, consideration should be given to how existing training programs can have non-technical skills training integrated into them, rather than developing a separate training program dedicated solely to non-technical skills.

Multidisciplinary and Team-Based Training

One of the most important considerations for non-technical skill development is that real-world performance in high-risk industries typically involves coordination and teamwork. While technical training is by nature largely discipline specific, this is not the case for non-technical skills. For instance, in the healthcare industry, the technical skills associated with performing surgery are very different from the technical skills involved in anaesthesia. However, patient safety often rests on the ability of the surgeon and the anaesthetist to communicate effectively with one another using a set of shared processes and skills.

Unlike technical expertise, non-technical skills are often critical to the coordination of work across distributed systems, where operators might not be working in the same location or have a single set of common goals. Aviation provides an excellent example here, where flight crew in the course of a normal flight must coordinate with air traffic control, with dispatch and load control, with maintenance engineers and with cabin crew, to name but a few.

Therefore, due to the critical nature of multidisciplinary work in high-risk industries, we need to consider how we design non-technical skills programs to replicate this aspect of real-world task performance.

First, consideration needs to be given to whether joint-training programs should be used, in which the different disciplines are brought together. In the early days of aviation CRM, the notion of team training between flight crew and cabin crew became the norm. This was in response to the catastrophic loss of Air Ontario Flight 1363 on 10 March 1989. This accident, which was caused by ice and snow accumulation on the wings, could have been averted if the concerns raised by a pilot flying as a passenger had been conveyed by the cabin crew to the flight crew. Today, many airlines develop training programs that involve a wide range of safety-critical staff, including flight crew, cabin crew, dispatchers and maintenance staff, particularly in seminar programs that focus on core enabling knowledge and safety attitudes.

EXAMPLE FROM PRACTICE

In high-acuity healthcare environments, such as the emergency department, a successful patient outcome depends on effective coordination between doctors from different specialties, nurses, and staff with other technical expertise, such as radiographers. In complex trauma cases such as those resulting from a high-speed motor vehicle accident, these tasks are coordinated in a high-pressure time-critical environment.

The most effective way to train the non-technical skills associated with coordinating the trauma team, such as communication and task management, is to use simulation involving the whole team. In this way, each of the different team members, with their unique expertise and unique roles, can develop their non-technical skills in an authentic team-based environment.

As will be explored in more detail throughout this book, simulation provides significant potential for skill development in the domains of non-technical skills. Ideally, joint training would also take place in the simulator, and this is achieved in a number of contexts, such as healthcare and process control training. However, sometimes it is not possible to bring together multidisciplinary teams, as is the case in aviation: pilots and air traffic controllers work for entirely different organisations. In these cases, it is necessary to consider how the multidisciplinary nature of work can be replicated in training to support task performance. Often, instructors may play the role of these agents, particularly in distributed systems, or they are woven into the training through scenario development and scripted events.

PRINCIPLE THREE: NON-TECHNICAL SKILLS TRAINING PROGRAMS SHOULD CONSIDER THE MULTIDISCIPLINARY AND TEAM-BASED NATURE OF WORK IN THEIR DESIGN

For non-technical skill training to be authentic and achieve an appropriate degree of transfer of training, consideration needs to be given to how the multidisciplinary nature of real work task performance is replicated in the training environment.

Generic and Abstracted or Tailored and Integrated

The scope of non-technical skills training programs lends considerable impact to the overall benefits derived from such programs with respect to skill development and sustainable changes in attitudes and behaviours.

Some training programs can be criticised for being too broad to be meaningful to specific operational tasks. Similarly, other programs can be criticised for being too specific to a particular situation, such that the skills developed are not easily transferred to other situations.

In general, the majority of non-technical skills training programs have historically adopted a broad scope and have introduced very generic domains of knowledge and skill that can then be applied to specific operational scenarios. This is reflected both in the content and design of training programs such as NOTECHS[4] and Anaesthetists' Non-Technical Skills (ANTS)[5] and in the design of assessment tools, as will be discussed in the following chapter. Indeed, it is possible to take these generic frameworks and apply them with little adaptation to other industrial environments. However, the dangers of this have been well demonstrated in the scientific literature.[3]

These generic programs can be criticised for the assumption that generic knowledge and skills at the domain level will easily transfer to the gamut of complex scenarios presented in day-to-day operations in high-risk industries. However, there is a body of evidence to highlight that this transfer is not easily achieved with traditional non-technical skills training programs.[6]

PRINCIPLE FOUR: CONSIDERATION SHOULD BE GIVEN TO THE SCOPE OF NON-TECHNICAL SKILLS TRAINING IN TERMS OF BEING GENERIC OR TASK SPECIFIC

The overall scope of a non-technical skills training program is an important consideration when designing training. Too broad, and the program is difficult to make meaningful to specific situations. Too narrow, and it is difficult for the training to be transferred to other situations. A blended design of generic principles and skills applied to a range of specific scenarios may provide the best of both worlds.

Attitudinal Change: Demonstrating Relevance and Importance

In the second chapter of this book, we explored the historical development of non-technical skills training programs and highlighted that the early models were often criticised for being 'psychobabble' and lacking direct relevance to a participant's everyday work. Therefore, it is absolutely critical that one of the first aspects of a successful training intervention is that is made clear to trainees how and why the training is important to their actual work.

For many industries, this process of demonstrating the relevance and importance of non-technical skills is seen as the very first stage of any non-technical skills training program, and often it is referred to as raising *awareness* of non-technical skills and human factors issues.[7] The awareness stage of non-technical skills training provides trainees with a common language for discussing and thinking about non-technical skills and establishes a clear understanding of the relationship between non-technical performance and safety.

A number of techniques can be used to demonstrate relevance and importance. The first is by reference to infamous examples of catastrophic failures within the relevant industry setting. The use of accident reports and reconstructions can easily demonstrate the role of the human element, and through careful facilitation, each of the areas of deficient non-technical skills can be drawn out. Another useful technique involves critical self-reflection, or reflection on the actions of peers or colleagues in an instance when something went wrong. When this is timed immediately after such an event, it is often termed a *critical incident debrief*. The process promotes a discussion of individual and team performance, identification of areas where performance could have been reviewed, and the development of strategies for subsequent enhanced performance.[8] Through the use of careful facilitation, the specific non-technical skills involved in the event can be explored in detail. An obvious alternative to this approach is to focus on positive events, where success has been the result of effective use of non-technical skills. Specific tools to facilitate such debriefing processes will be explored in the next chapter.

Demonstrating the relevance and importance of non-technical skills is also critical for the development of attitudes likely to promote appropriate safety behaviours in the workplace.[7] Early in the history of non-technical skills programs, the link between an operator's attitudes and safety performance was examined in detail. It was held that unsafe attitudes and inappropriate cognitions were significant contributing factors to the role of human error in accident causation.[9]

Therefore, from the outset, aviation CRM programs set out to change attitudes towards communication and teamwork on the flight deck.[10] This was originally referred to as an overarching aim of *indoctrination*, which formed a major component of early CRM training programs.

Due in part to the fact that cognitive and social psychologists were actively involved in training program design, as examined in Chapter 2, particular emphasis was placed on measuring trainees' attitudes and the attitudinal change associated with participating in non-technical skills training. Having trainees complete a safety attitudes questionnaire as part of the training program can help establish the relationship between non-technical skills and safe and efficient operations, and can also model appropriate safety attitudes and behaviours.

Simple tools such as the Safety Attitudes Questionnaire (SAQ)[11] and the Flight Management Attitudes and Safety Survey (FMASS)[12] have been developed to assess safety attitudes and have been adapted for use in many industries. Moreover, participation in non-technical skills training programs that focus on awareness raising and basic knowledge development have been shown to result in positive attitudinal change.[13,14]

PRINCIPLE FIVE: ATTITUDINAL CHANGE IS AN IMPORTANT OUTCOME OF NON-TECHNICAL SKILLS TRAINING PROGRAMS

Case studies of well-known accidents, as well as critical incident debriefs, are two effective ways to demonstrate the relevance and importance of non-technical skills in safe and efficient operations. Having trainees complete a safety attitudes questionnaire as part of the training program can help establish the relationship between non-technical skills and safe and efficient operations.

Core Knowledge Development

In the previous chapter, we explored a number of principles of adult learning, which emphasise that learning is most effective when a scaffolding of basic knowledge is developed first, on which further knowledge and skill development can then be built. This knowledge can be referred to as *enabling knowledge*, and it is essential to enhanced non-technical performance in individuals and teams.

PRINCIPLE SIX: CORE KNOWLEDGE NEEDS TO BE BUILT AS A FOUNDATION FOR SKILL DEVELOPMENT

Developing a core body of knowledge about non-technical skills is essential as a foundation upon which skills and attitudes can be formed.

Classroom-Based Training

There are a number of different ways in which knowledge development can be facilitated, with the history of non-technical skills training programs demonstrating that the classroom-based seminar is by far the most popular training mode for knowledge development. Many classroom-based non-technical skills training programs adopt a seminar format, in which key topics are explored. Often, these adopt a largely didactic teaching format, with each key topic being introduced through a summary of key points being presented by the facilitator, and usually with the help of the ubiquitous PowerPoint presentation.

This 'chalk and talk' approach to instruction does have its merits, and the model of the expert passing on key bite-sized pieces of core knowledge to the trainee is age-old. It is an efficient way to summarise key points and distil a topic into its most elementary form for trainees to build foundational knowledge. However, on its own, this mode of instruction can only assist in the development of basic awareness and understanding of key points. To develop real non-technical skills, much more participatory modes of learning are required.

PRINCIPLE SEVEN: THERE ARE SIGNIFICANT LIMITATIONS TO THE USE OF CLASSROOM AND eLEARNING MODES IN NON-TECHNICAL SKILLS TRAINING PROGRAMS

A variety of classroom-based and eLearning modes can be used for the development of core enabling knowledge in non-technical skills training programs – but these modes often make it difficult to create the experiential forms of learning that are required for skill development.

Thankfully, without too much imagination, the classroom can also become the venue for more advanced forms of instruction and can, to some degree, even be used to facilitate experiential form of learning. In particular, the use of the case study technique and role-play simulations can be used to great effect in the classroom. These forms of instruction will be explored later in this chapter.

The Use of eLearning Platforms

The term *eLearning* is used here to capture the various forms of technology-based training programs, from internet-based training delivered through a learning management system to individualised forms of computer-based training (CBT) delivered within a training department. The many forms of eLearning available today provide efficient and flexible forms of training for the development of core enabling knowledge in non-technical skills. Moreover, the use of multimedia can create engaging learning environments that meet the needs of trainees with different learning styles, and the use of multimedia can assist in modelling the complex systems seen in high-risk industries.

Since the 1990s, the use of forms of eLearning has become important for training delivery in many high-risk industries. Given the rapid evolution of the internet and computer-based training in general, there are now a significant number of benefits that can be accessed through the use of technology-based learning platforms.

Of these benefits, perhaps the most important relate to the approach to learning facilitated through eLearning systems. For instance, eLearning offers considerable *flexibility* in terms of when and where a learner engages with the training program. Moreover, through the use of multimedia, the training can be tailored to the different *learning styles* of individual trainees. These aspects, in turn, can realise a reduction in the cost of training programs.

Another set of benefits relates to what can be achieved through the medium of eLearning. For instance, eLearning platforms are able to *model complex systems* such as parts of human physiology, an aircraft's hydraulic systems, or a high-voltage electricity network. Further, they are able to re-create normal and non-normal system states, which can contribute significantly to the development of a trainee's situation awareness and decision-making skills.[15]

Today, most forms of eLearning can be realised though internet-based platforms, which further increase the flexibility and reduce the costs associated with training delivery. Moreover, the use of internet-based eLearning platforms enables the delivery of training through an integrated learning management system, whereby course content can be tailored to individuals, assessment can be delivered, and training records can be maintained within a single system.[16]

Case Study Techniques

Case study techniques are a powerful form of learning in non-technical skills programs. Although they require thorough preparation and skilful facilitation, they can lead to the development of complex problem-solving skills that can transfer to the work environment.

The role of storytelling has always been a critical part of knowledge development. From the use of parables and folklore to shape children's understanding of the world, through to the formal case study methodologies, narrative is a critical component of sharing knowledge and developing new understandings.

Perhaps the best-documented case study technique is the case method developed by the Harvard Business School. Although developed for use in developing skills in business administration, such as financial management and human resources, it is easily adapted to the development of awareness and knowledge relating to non-technical skills.

The case method involves the presentation of a pre-defined scenario or event to trainees. This may involve a range of media, including written materials, videos, images and diagrams. The scenario is presented in the form of a problem, which the trainees must attempt to solve. This problem-solving process typically involves active discussion by the group of trainees, which is facilitated by questions and challenges from the instructor. More details around effective facilitation techniques will be explored in the following chapter.

Training programs that use a case study approach draw on a number of aspects of adult learning theory described in the previous chapter. A well-presented case can be so engaging that it activates affective and cognitive forms of learning and ensures relevance through the use of real-world and industry-specific scenarios. Also, it requires trainees to draw on and integrate basic forms of knowledge and apply these to a real-world situation.

The way in which the use of case study techniques results in transfer of training is through a process termed *case-based reasoning*. Case-based reasoning is a form of analogical reasoning when a novel, and often ill-defined, problem is encountered. The problem is solved by reference to previously encountered cases that exhibit some form of similarity.[17] This means that operators, having had to deal with solving a complex problem within a case study presented during training, are better equipped to deal with similar or analogous problems in the real work environment.

Recent research has highlighted different learning outcomes associated with cases dealing with a situation that ended in either success or failure. While, intuitively, cases that result in a successful outcome might be thought to result in better learning outcomes, as they provide exemplars of actions that result in optimal outcomes, the evidence suggests otherwise. Presenting cases that result in failure has been demonstrated to result in enhanced learning outcomes, due to the fact that learners must engage more deeply with the case to determine exactly what the causes of the failure were.[18]

Video-Based Case Study Techniques

Short films or videos have been used in the training of non-technical skills for many decades and across many domains. There are a number of ways in which video-based case studies can be used and, if designed carefully, can actually contribute to skill development. The use of video is not simply about increasing the enjoyment a trainee has during training. Rather, a well-designed video can actually facilitate non-technical skill development through critical reflection, analogical reasoning and behavioural modelling.

Awareness Raising: In their simplest use, video-based case studies provide an engaging way to raise awareness of the role of non-technical skills and demonstrate their relevance and importance. The popularity of documentary series such as *Air Crash Investigation* and *Seconds from Disaster* highlights that video-based case studies are a potentially powerful tool.

Discussion Starters: In the domain of medical education, the term *trigger films* was first coined in the 1980s to describe short movie vignettes used to present a scenario that could then be subjected to critical reflection and discussion in the training environment.[19] Also, short videos featured heavily in early aviation CRM courses, and those early videos are sometimes quite extraordinary to watch today!

Behavioural Modelling: A more sophisticated use of video-based case studies involves their use as templates for behavioural modelling. Along the lines of the axiom 'see one, do one, teach one', observing expert performance is an important facet of learning, as described in the previous chapter. With careful design, scripting and production, a video can be created to demonstrate complex aspects of non-technical performance such as graded assertiveness or workload management.

Non-Technical Skill Development and Rehearsal

Role-Play Simulation

From the very early CRM programs, role-play has been used to create a bridge between the classroom and the real-world environment of the workplace. To achieve a less didactic and more active form of learning, role-play often involves replicating aspects of real-world activities in the classroom. Role-play is a simple and cost-effective form of simulation, which can contribute significantly to the development of non-technical skills. Role-play enables more experiential forms of learning to be achieved even in the classroom environment. Reluctance to participate in role-play can be overcome through the careful design of realistic scenarios that represent a complex and ill-defined problem in real-world operations.

Sometimes, role-play is seen as a childlike training technique and can meet considerable resistance. However, it is actually a highly valid form of low-fidelity simulation and can be used to great effect in bringing more experiential forms of learning to the classroom.

While there are many forms of role-play, the one most used in the training of non-technical skills involves presenting a hypothetical situation to a group and selecting individuals to assume roles within that scenario, which is then 'acted out' to facilitate skill development.[20] As the role-play exercise unfolds, trainees must react to the scenario presented as well as the actions of other participants in the role-play. Following the role-play, facilitated discussion can then be used to provide feedback and explore aspects of good and poor performance.[10] In this way, many of the benefits of simulation-based training can be realised without the expense.

The use of role-play in non-technical skills training has a considerable number of benefits. First, it is a simple and low-cost form of

Table 4.1 Guidelines for Developing Role-Play Exercises

GUIDELINE	DESCRIPTION
Identify training needs	As with any form of training, the learning activities must be designed explicitly to develop pre-defined needs. The role and processes of training needs analysis will be explored in more detail in the following chapter, which deals explicitly with Instructional Systems Design.
Use subject matter experts	The scenarios used in role-play must be realistic and elicit the knowledge and skills required for real-world performance. Subject matter experts are required to ensure that the scenarios have high levels of face validity.
Provide structure to the scenario	The scenario used in the role-play must be supported, such that it is able to naturally unfold and events are cued along a timeline. Strategies such as presenting cue cards to the participants when a new event happens are useful.
Ensure that the roles of participants are well defined	Each participant should be provided with clear instructions as to their role in the scenario, which may include cues as to the unfolding scenario, the expected responses to events and other participants' actions. Usually, the participant is only given information about their own role, and not those of others, to ensure that responses are spontaneous.
Create an atmosphere for the role-play	The use of low-fidelity props, such as a mock-up cockpit or control room, can significantly enhance the face validity of the scenario.
Provide opportunities for practice	Ideally, all participants in the training session should be able to assume each of the different roles and be given opportunities to practice. However, this may not be achievable given time constraints and the number of trainees. Even observation has benefits in terms of behavioural modelling and skill development.
Provide guidance for the facilitator	The facilitator must be provided with guidance with respect to managing the role-play scenario and also their expected roles in terms of providing feedback, facilitating debriefing, and other forms of facilitation.

Source: Beard, R.L. et al., *Int. J. Aviat. Psychol.*, 5(2), 131, 1995.

simulation, in which experiential forms of learning and rehearsal can be achieved with a low cost and therefore can provide a high return on the investment of the training budget. Second, it can bring these experiential forms of learning out of the relative privacy of the higher-fidelity simulation and enable a larger group to participate in observational forms of learning such as behavioural modelling and critical reflection.

Some very good guidance has been developed for the design of role-play exercises in non-technical skills training programs within the scientific literature. Table 4.1 provides a brief summary of some of the important principles for the design of effective role-play exercises.

Simulation and Training Non-Technical Skills

Across a wide array of industries, simulation has become a critical component of training programs and has opened up amazing opportunities for the development of both technical and non-technical skills. Simulation has reformed training in high-risk industries such as aviation, rail, healthcare and nuclear power generation, to name but a few. Today, because of simulation, training in these contexts is now more realistic, safe, cost-effective and flexible than ever before.[21] Simulation enables the rich complexity of real-world operations to be brought into a highly controlled training environment. It is used extensively in the development of technical skills and is being used more and more for the development of non-technical skills.

Simulation and Learning

There are a number of characteristics of simulation that can be used to enhance non-technical skill development and provide benefits that cannot be otherwise realised. Recent empirical evaluations have highlighted that simulation use is an important aspect of non-technical skill development and may have an additive effect whereby enhanced knowledge, skills and attitudes can be generated through the combined exposure to simulation and traditional didactic forms of training.[22] Each of these unique characteristics of simulation-based training maps onto an aspect of adult learning theory discussed in the previous chapter, and they can be briefly summarised as follows.

The first of these is that simulation can create an *authentic* learning environment without the contextual constraints, such as time and costs, associated with using real equipment and working environments in training. Furthermore, simulation can contextualise learning in the same social context of real-world operations, allowing the social negotiation of new knowledge and skill.

Second, simulation can allow skill development to take place in an environment that is *safe*, where there is no risk to the equipment (such as an aeroplane, a train or a nuclear power plant) or the people involved in the training (such as the trainees and instructors) as well as the general public. Perhaps even more importantly, this idea of safety extends to the *psychological safety* of participants. Simulation is an environment where sub-optimal performance and error can occur without fear of negative consequence. Indeed, if the instructional environment is established appropriately, participants are to expect that errors will occur, and these represent powerful learning opportunities that would be otherwise inaccessible in traditional forms of training.

Third, simulation can enable the use of instructional techniques that are much more difficult to achieve in the real world. Of these, the fluidity of time in simulation offers a number of benefits. For instance, the ability to 'freeze' reality and pause the simulation allows moments of reflection and debrief, which could not occur if events kept unfolding in real time. Further, simulation allows practice and rehearsal much more easily than training in real-world environments, as the training session can jump forwards or backwards to different points in a situation as it unfolds, which is quite different from the organic 'what you see is what you get' nature of training in real-world contexts.

Finally, simulation enables highly controlled and highly scripted scenarios in which any number of aspects of the work situation can be manipulated. For instance, a scenario aimed at developing *decision-making skills* can generate a circumstance where there is no single correct answer, and trainees must draw on complex reasoning and risk-based judgements. Similarly, a scenario aimed at developing *task management skills* can introduce a number of concurrent competing demands to create a high-workload situation, or it can introduce subtle forms of distraction that must be effectively managed by the individual or team. As will be discussed in the following sections and chapters of this book, carefully considered scenario design is one of the most important features of a successful non-technical skills training program.

PRINCIPLE EIGHT: SIMULATION PRESENTS A HIGHLY EFFECTIVE TRAINING MODE FOR THE DEVELOPMENT OF NON-TECHNICAL SKILLS

Simulation enables the rich complexity of real-world operations to be brought into a highly controlled training environment. It is used extensively in the development of technical skills and is being used more and more for the development of non-technical skills.

Simulation Fidelity

There is often the misconception that simulation must involve expensive full-scale reproductions of the actual work environment, such as the flight deck, the control room or the operating theatre. Such simulators are referred to as *high-fidelity simulators*. The universal appeal of modern technology has perhaps led to an over-use of high-fidelity simulation in training across high-risk industries, and the reduction of fidelity to a mono-dimensional construct relating to the fidelity of the technology involved. In turn, this has led to a focus on what the simulator can do, rather than how it is used to maximise learning and the transfer of training.[21] There are multiple dimensions of fidelity, and using the most expensive ultra-high-fidelity representation of the actual work environment does not guarantee better learning outcomes. Table 4.2 provides an overview of the many dimensions of simulation fidelity.[23,24]

Another concept of fidelity, and one that is critical to the development of non-technical skills training programs, is *operational* or task fidelity.[24] This aspect of fidelity relates to the degree to which the simulation reflects actual real-world events and the context in which they occur. Closely aligned to the concept of authentic learning introduced in the previous chapter, operational fidelity is not necessarily related to technological fidelity; rather, it requires careful consideration of the context in which work occurs and the types of situations that demand high levels of non-technical performance. This concept of operational

Table 4.2 Dimensions of Fidelity

DIMENSION	DESCRIPTION
Physical	The degree to which the physical characteristics of the simulator match those of the real technology and task.
Engineering/ technological	The degree to which the simulator is built in the same way as the real machine or equipment.
Environmental	The degree to which the simulator can re-create the range of environmental characteristics within which work is performed in the real world.
Psychological	The degree to which trainees view the simulation as realistic.
Task	The degree to which the simulation asks trainees to perform real-world tasks.
Operational	The degree to which the simulator can re-create the complex and messy nature of real-world operations.

Sources: Stanton, N.A., *Human Factors in Nuclear Safety*, London, Taylor and Francis, 1996 and Thomas, M.J.W., *Proceedings of SimTecT2003: Simulation Conference*, Adelaide, Australia, Simulation Industry Association of Australia, 2003.

fidelity leads us to the critical aspect of simulation-based training: *scenario design*.

Scenario Design for Simulation-Based Training

Beyond the hardware of the simulator, the most important factor that influences the educational benefit of simulation-based training relates to the scenarios that are used in specific training interventions. The scenario is the very heart of simulation-based training and describes in detail a series of events as they unfold.

The first principle of effective scenario design, as with any form of training, is to first identify the learning objectives of that training and to build the scenario with these outcomes in mind.[25] With the training of non-technical skills, this process is relatively straightforward if an evidence-based approach is used. Drawing on incident and accident reports is often a useful approach to begin the process of scenario design. For instance, if decision-making skills are to be the focus of the scenario, a review of incidents within the organisation that has highlighted sub-optimal decision-making can provide excellent guidance for creating the task environment and relevant contextual factors for an appropriate scenario.

A particularly effective approach to simulation scenario design is the event-based approach to training (EBAT).[26,27] This approach is designed to directly link the content of the scenario with the competencies that are to be trained. First, the approach ensures that a training needs analysis has been undertaken and that competencies have been clearly specified. Next, events are identified that will elicit performance in the skill domains identified in the training needs analysis and competency specification. Additional considerations for scenario design are provided in Table 4.3.

Overarching Instructional Design Considerations

Overall, the effective use of simulation in training and assessing non-technical skills relates to the appropriate application of adult learning theory and the overall instructional design processes used in the development of the non-technical skills training program. These will be explored in detail in the following chapters.

Table 4.3 Considerations for Scenario Design

DIMENSION	DESCRIPTION
Events	As per the EBAT approach, events are identified that will elicit performance of trainees in the skill domains identified in the training needs analysis and competency specification.
Context	The workplace context in which the events would normally take place is described and re-created with an appropriate degree of fidelity for the training situation.
Equipment and resources	The required equipment and resources, such as patient notes, a flight plan or job safety analysis forms, are identified and made available for the scenario.
Script	A script is produced that includes the timing of events, any scripted interactions for the instructor, actors or confederates, and guidance as to what to do in the event that the scenario takes different paths as it unfolds.
Rehearsal	Prior to being used in training, the scenario should be rehearsed to ensure smooth running and that any unanticipated problems are resolved.

Skill Development and Rehearsal on the Job

In many high-risk industries, such as healthcare, aviation, rail transport and utilities, a large proportion of knowledge and skill development takes place through formal training programs. However, even in these environments, non-technical skills continue to be developed on the job.

There are a number of ways in which on-going skill development takes place, either through formal training interventions embedded within workplace activities, or informally through behavioural modelling.

PRINCIPLE NINE: NON-TECHNICAL SKILLS TRAINING NEED NOT BE RESTRICTED ONLY TO FORMAL TRAINING PROGRAMS

A number of informal training modes, such as mentoring and coaching and other forms of on-the-job training, can be used effectively in non-technical skills development.

Mentoring, Coaching and Apprenticeship Models

For centuries, on-the-job training was commonplace, and indeed, formal vocational training programs outside the workplace have only relatively recently replaced aspects of the apprenticeship model. Mentoring and coaching involve the observation and monitoring of trainees carrying out work-related activities and providing feedback to facilitate learning. In this on-the-job training model, experts provide hints, clues and tricks of the trade to assist learners in achieving skill development.[28] Forms of training embedded within the apprenticeship models of skill development offer excellent opportunities for on-going development of non-technical skills. However, this is best done within a formal and planned structure, rather than in an *ad hoc* manner.

While not perhaps formally labelled as coaching *per se*, this model of on-the-job skill development has been the dominant paradigm used in medical education of specialists such as anaesthetists and surgeons. The medical apprenticeship model is interesting in a number of respects, not the least of which is the way in which technical and non-technical aspects of skill development are integrated. In this model, non-technical skills such as diagnosis and clinical reasoning are developed alongside advanced technical knowledge relating to physiology and pathology.

In other domains, such as air traffic control and rail, on-the-job training is also commonplace, and consideration in these domains needs to be given to how non-technical skills can be integrated into these forms of training. For instance, it has been suggested that more emphasis needs to be placed on integrating formal team training exercises into domains where teamwork is critical but often not explicitly trained during on-the-job training.[29]

Research has also highlighted the important role that on-going coaching plays in sustaining skill development after other forms of non-technical training interventions. For instance, transfer of training, skill maintenance and further skill development after a simulator-based non-technical skills training program relies on the on-the-job reinforcement of those skills through mentoring and coaching.[30]

EXAMPLE FROM PRACTICE: ASSOCIATED NON-TECHNICAL SKILLS IN MINING

Worldwide, the mineral and extractive industries present a high-risk workplace, where sub-optimal performance in non-technical skill domains such as situation awareness and communication can all too easily lead to fatal injury. In Australia, an innovative program called Associated Non-Technical Skills has been developed, which includes interventions for coaching non-technical skills in an on-the-job training format.[31]

In this program, potential coaches are trained both in the core domains of non-technical skills and in a process of on-the-job observation, feedback and skill development. Coaches then work with groups of workers on the job, and through cycles of observation and facilitated discussion, assist workers in enhancing their non-technical skills. The domains for coaching within the program include (1) Communication; (2) Situation Awareness; (3) Decision-Making; (4) Leadership; and (5) Teamwork.

Behavioural Modelling

Another, less formal, way in which non-technical skills are developed is through naturalistic forms of learning where the behaviour of an expert is observed and subsequently used as a template for the trainee's behaviour. Linked to the concepts of the reflective practitioner and cognitive apprenticeships explored in the previous chapter, these forms of learning are not restricted to overt behaviours, but include cognitive activity, which is critical for non-technical skill development. The modelling of cognitive behaviours can be facilitated by the expert speaking aloud their thinking process, or by the trainee asking questions about the thinking behind specific actions.

Organic forms of on-the-job training, such as behavioural modelling, can form effective elements of an overall training program for non-technical skills development. Unfortunately, the modelling of inappropriate behaviours is also a common form of learning, and in some settings, has led to the perpetuation of cultures of bullying in industries such as healthcare.[32]

Briefing and Debriefing

Briefing is the process of establishing the context for the training event. From the perspective of adult learning theory, briefing serves the purpose of activation and priming for learning, as discussed in the previous chapter.[33] Briefing is a critical part of any non-technical skills training programs and serves a number of functions associated with knowledge and skill development, such as priming existing knowledge, mental simulation of behaviours prior to purposeful practice, and setting the expectations for the training session.

With respect to the simulation-based training of non-technical skills, briefing also serves a number of additional purposes. First, briefing allows the instructor to set out the context of the simulator scenario, and adds to realism and user engagement by orientating the learner in their mind to the context of the scenario prior to entering the simulator. In short, it is a critical ingredient in the 'magic' of the simulation. Second, briefing should set out the roles and expectations of both trainee and instructor.[34] This is especially the case where the instructor may take on several peripheral roles in the simulation (such as an air traffic controller) or may plan to 'freeze' the simulation scenario to prompt critical reflection or provide feedback.

Perhaps most importantly, briefing prior to non-technical skills training is critical in establishing the environment in which trainees can feel safe to develop new skills and try new strategies. The ideal formats of non-technical skills training, such as simulation-based training, necessarily involve individuals attempting to manage complex situations in front of their peers. To maximise the learning outcome and minimise any negative or adverse impacts of the training experience, a psychologically safe learning environment must be created. In the medical domain, this has been formalised with the term *safe container* for non-technical skills development, such that frank evaluation and reflection on performance can take place.[35]

In addition to the briefing provided by the instructor, it has also been suggested that in industries where pre-task briefing is undertaken, such as the pre-departure brief in aviation, team time out in the peri-operative environment in healthcare, or take five in construction, teams should also undertake that briefing prior to any simulation-based training.[25]

Debriefing is a critical element of any non-technical skills training program. However, since it relates equally to the assessment of the non-technical performance of an individual or team, debriefing will be explored in significant detail in the following chapter.

PRINCIPLE TEN: BRIEFING AND DEBRIEFING ARE ESSENTIAL COMPONENTS OF ANY NON-TECHNICAL SKILLS TRAINING SESSION

The processes of briefing and debriefing are critical in maximising the learning benefits of a non-technical skills training session. Skills in the effective facilitation of briefing and debriefing should form the basis of training instructors for non-technical skills training programs.

References

1. Civil Aviation Authority. (2014). *Flight-Crew Human Factors Handbook: CAP 737*. London, UK: Civil Aviation Authority UK.
2. O'Connor, P., Campbell, J., Newon, J., Melton, J., Salas, E., and Wilson, K. A. (2008). Crew resource management training effectiveness: A meta-analysis and some critical needs. *International Journal of Aviation Psychology, 18*(4), 353–368.
3. Flin, R., and Maran, N. (2004). Identifying and training non-technical skills for teams in acute medicine. *Quality and Safety in Health Care, 13*(suppl 1), i80–i84.
4. Flin, R., Martin, L., Goeters, K.-M., Hormann, H., Amalberti, R., Valot, C., and Nijhuis, H. (2003). Development of the NOTECHS (Non-Technical Skills) system for assessing pilots' CRM skills. *Human Factors and Aerospace Safety, 3*, 97–120.
5. Fletcher, G., Flin, R., McGeorge, P., Glavin, R. J., Maran, N. J., and Patey, R. (2003). Anaesthetist's non-technical skills (ANTS): Evaluation of a behavioural marker system. *British Journal of Anaesthesia, 90*(5), 580–588.
6. Zausig, Y., Grube, C., Boeker-Blum, T., Busch, C., Bayer, Y., Sinner, B., Zink, W., Schaper, N., and Graf, B. (2009). Inefficacy of simulator-based training on anaesthesiologists' non-technical skills. *Acta Anaesthesiologica Scandinavica, 53*(5), 611–619.

7. Civil Aviation Safety Authority. (2011). *Non-Technical Skills Training and Assessment for Regular Public Transport Operations CAAP SMS-3(1)*. Canberra, Australia: Civil Aviation Safety Authority.
8. Salas, E., Klein, C., King, H., Salisbury, M., Augenstein, J. S., Birnbach, D. J., Robinson, D. W., and Upshaw, C. (2008). Debriefing medical teams: 12 evidence-based best practices and tips. *The Joint Commission Journal on Quality and Patient Safety, 34*(9), 518–527.
9. Helmreich, R. L., Wilhelm, J. A., Klinect, J. R., and Merritt, A. C. (2001). Culture, error and crew resource management. In E. Salas, C. A. Bowers and E. Edens (Eds.), *Improving Teamwork in Organizations* (pp. 305–331). Mahwah, NJ: Lawrence Erlbaum Associates.
10. Beard, R. L., Salas, E., and Prince, C. (1995). Enhancing transfer of training: Using role-play to foster teamwork in the cockpit. *International Journal of Aviation Psychology, 5*(2), 131.
11. Sexton, J. B., Helmreich, R. L., Neilands, T. B., Rowan, K., Vella, K., Boyden, J., Roberts, P. R., and Thomas, E. J. (2006). The safety attitudes questionnaire: Psychometric properties, benchmarking data, and emerging research. *BMC Health Services Research, 6*(44), 44–54.
12. Sexton, B. J., Helmreich, R. L., Wilhelm, J. A., Merritt, A. C., and Klinect, J. R. (2001). *Flight management attitudes and safety survey (FMASS): A shortened version of the FMAQ*. University of Texas at Austin Human Factors Research Project, Technical Report 01-01.
13. Helmreich, R. L., and Wilhelm, J. A. (1991). Outcomes of crew resource management training. *International Journal of Aviation Psychology, 1*(4), 287–300.
14. Salas, E., Burke, C. S., Bowers, C. A., and Wilson, K. A. (2001). Team training in the skies: Does crew resource management (CRM) training work? *Human Factors, 43*(4), 641–674.
15. Wiggins, M. W. (2003). Computer-based approaches in aviation education. In I. M. A. Henely (Ed.), *Aviation Education and Training: Adult Learning Principles and Teaching Strategies* (pp. 316–345). Aldershot, UK: Ashgate Publishing.
16. Thomas, M. J. W. (2003). Internet-based education and training: Using new technologies to facilitate instruction and enhance learning. In I. M. A. Henely (Ed.), *Aviation Education and Training: Adult Learning Principles and Teaching Strategies* (pp. 346–366). Aldershot, UK: Ashgate Publishing.
17. Kolodner, J. (2014). *Case-Based Reasoning*. San Mateo, CA: Morgan Kaufmann.
18. Tawfik, A., and Jonassen, D. (2013). The effects of successful versus failure-based cases on argumentation while solving decision-making problems. *Educational Technology Research and Development, 61*(3), 385–406.
19. Alroy, G., and Ber, R. (1982). Doctor-patient relationship and the medical student: The use of trigger films. *Academic Medicine, 57*(4), 334–336.
20. Killen, R. (2006). *Effective Teaching Strategies: Lessons from Research and Practice*. South Melbourne, Australia: Cengage Learning Australia.

21. Salas, E., Bowers, C. A., and Rhodenizer, L. (1998). It is not how much you have but how you use it: Toward a rational use of simulation to support aviation training. *International Journal of Aviation Psychology, 8*(3), 197.
22. Shapiro, M., Morey, J., Small, S., Langford, V., Kaylor, C., Jagminas, L., Suner, S., Salisbury, M., Simon, R., and Jay, G. (2004). Simulation based teamwork training for emergency department staff: Does it improve clinical team performance when added to an existing didactic teamwork curriculum? *Quality and Safety in Health Care, 13*(6), 417–421.
23. Stanton, N. A. (1996). Simulators: A review of research and practice. In N. A. Stanton (Ed.), *Human Factors in Nuclear Safety* (pp. 114–137). London: Taylor and Francis.
24. Thomas, M. J. W. (2003). Operational fidelity in simulation-based training: The use of data from threat and error management analysis in instructional systems design. In *Proceedings of SimTecT2003: Simulation Conference* (pp. 91–95). Adelaide, Australia: Simulation Industry Association of Australia.
25. Prince, C., Oser, R., Salas, E., and Woodruff, W. (1993). Increasing hits and reducing misses in CRM/LOS scenarios: Guidelines for simulator scenario development. *The International Journal of Aviation Psychology, 3*(1), 69–82.
26. Rosen, M. A., Salas, E., Wu, T. S., Silvestri, S., Lazzara, E. H., Lyons, R., Weaver, S. J., and King, H. B. (2008). Promoting teamwork: An event-based approach to simulation-based teamwork training for emergency medicine residents. *Academic Emergency Medicine, 15*(11), 1190–1198.
27. Fowlkes, J., Dwyer, D. J., Oser, R. L., and Salas, E. (1998). Event-based approach to training (EBAT). *The International Journal of Aviation Psychology, 8*(3), 209–221.
28. Billett, S. (1994). Situating learning in the workplace-having another look at apprenticeships. *Industrial and Commercial Training, 26*(11), 9–16.
29. Ellis, K. H., and McDonell, P. (2003). On-the-job training in air traffic control. In I. M. A. Henely (Ed.), *Aviation Education and Training* (pp. 367–387). Aldershot, UK: Ashgate Publishing.
30. Salas, E., DiazGranados, D., Weaver, S. J., and King, H. (2008). Does team training work? Principles for health care. *Academic Emergency Medicine, 15*(11), 1002–1009.
31. Shaw, A., Oakman, J., Thomas, M. J. W., Blewett, V., Aickin, C., Stiller, L., and Riley, D. (2014). *Associated Non Technical Skills: Action Learning Program Facilitators Resource*. Sydney, NSW: Department of Industry, Resources and Energy.
32. Berk, R. (2009). Derogatory and cynical humour in clinical teaching and the workplace: The need for professionalism. *Medical Education, 43*(1), 7–9.
33. Merrill, M. D. (2002). First principles of instruction. *Educational Technology Research and Development, 50*(3), 43–59.

34. Federal Aviation Administration. (2015). *Advisory Circular 120-35D: Flightcrew Member Line Operational Simulations: Line-Oriented Flight Training, Special Purpose Operational Training, Line Operational Evaluation*. Washington, DC: US Department of Transportation.
35. Rudolph, J. W., Raemer, D. B., and Simon, R. (2014). Establishing a safe container for learning in simulation: The role of the presimulation briefing. *Simulation in Healthcare, 9*(6), 339–349.

5
Principles of Assessing Non-Technical Skills

Introduction

Assessment is a critical component of any training program. At its most basic, assessment can be described as the process of measuring a trainee's knowledge, skills or attitudes. Assessment serves a number of different purposes in a training program. As introduced in previous chapters, there are two main forms of assessment: *formative* assessment and *summative* assessment. While summative assessment takes place at the completion of a training program and assesses the degree to which knowledge and skills have been acquired by the trainee, formative assessment takes place during the course of a training program and is typically used to measure and give feedback on learning progress and to tailor additional training interventions.

One of the challenges facing the assessment of non-technical skills is that they are to some degree harder to measure than technical skills. This is particularly the case for the non-technical skills within the cognitive domain, such as situation awareness and decision-making. With these non-technical skills, traditional approaches to assessment might only be able to assess the outcomes of non-technical performance and not easily access the processes that give rise to outcomes.

This chapter will introduce a set of principles that can be used in the design of assessment programs for non-technical skills. The principles introduced in this chapter are as follows:

- Principle One: Assessing core knowledge is an important first stage of non-technical skills assessment, but it is only the first step.

- Principle Two: Assessing non-technical skills should adopt a competency-based approach.
- Principle Three: The use of a behavioural marker system is an effective, and widely used, technique for assessing non-technical skills.
- Principle Four: A balance needs to be achieved between generic sets of behavioural markers and very specific check-lists of desirable/undesirable actions in a given scenario when assessing non-technical skills.
- Principle Five: An integrated approach to assessing non-technical skills, in which technical aspects of performance are also evaluated, is highly desirable.
- Principle Six: As a critical determinant of safety in high-risk industries, poor non-technical performance should be reason enough for an individual to be deemed not competent.
- Principle Seven: Instructor training, calibration and inter-rater reliability are critical aspects of the effective assessment of non-technical skills.
- Principle Eight: A set of biases that can negatively impact on the assessment of non-technical skills must be identified and rectified on a regular basis.
- Principle Nine: Self-assessment of non-technical skills should be approached with caution, as trainees often lack insight into their own performance.
- Principle Ten: Facilitated debrief of performance is a critical element of assessing non-technical skills and maintains a continuous cycle of training, assessment and enhanced performance.

Assessing Core Enabling Knowledge

The first aspect of assessment relevant to non-technical skills training programs relates to assessing the development of the core enabling knowledge required for optimal performance of non-technical skills in the complex real-world environment.

Unfortunately, many training programs choose to rely on very simple test forms such as multiple choice, in which only basic declarative

Table 5.1 Summary of Techniques for Assessment of Knowledge

TECHNIQUE	ASSESSMENT TARGET
True – False quiz	Declarative knowledge
Multiple choice	Declarative and procedural knowledge
Short answer	Problem-based
Longer format (essay)	More complex problems, synthesis of core concepts
Oral	More complex problems, synthesis of core concepts

knowledge (statements of fact) is assessed. However, it is preferable to assess the degree of knowledge developed in a non-technical skills training program using a variety of techniques. Table 5.1 summarises the techniques that are often used for the assessment of core enabling knowledge.

A non-technical skills training program will often structure assessment tasks to confirm that enabling knowledge has been developed prior to sequencing into dedicated forms of skill development, deliberate practice and then competency-based assessment in real task situations. If forms of eLearning are used, assessment of enabling knowledge is sometimes integrated into these packages or included as a component of pre-training reading to maximise the practical training opportunities in the face-to-face environment of the simulator or on-the-job training.

PRINCIPLE ONE: ASSESSING CORE KNOWLEDGE IS AN IMPORTANT FIRST STAGE OF NON-TECHNICAL SKILLS ASSESSMENT, BUT IT IS ONLY THE FIRST STEP

Assessing core knowledge of non-technical skills is a critical first step, as it ensures that a theoretical understanding of the skill area is established as a foundation for effective skill development and task performance. However, knowledge-based assessment techniques by themselves are not sufficient in an effective non-technical skills training program.

Competency-Based Assessment of Non-Technical Skills

The most appropriate theoretical framework for the assessment of non-technical skills is a competency-based approach. As described in Chapter 3, competency-based training is an approach that focusses directly on the specified knowledge, skills and attitudes that enable task performance. It focusses on developing and demonstrating that someone is competent to perform safety-critical work.

There is a considerable body of research exploring effective competency-based assessment. The major principles are as follows:

- Assessment focuses on the combined *demonstration of knowledge, skills and attitudes* and not simply a theoretical understanding of the non-technical skills domain.
- Assessment is undertaken while conducting *actual work tasks* either with the use of simulation or in the real-world work environment.
- Assessment is *criterion-based*, which means that specific predetermined criteria for demonstration of competence are used to identify whether a trainee is competent or not yet competent in the specific non-technical skill that is the focus of assessment.

EXAMPLE IN PRACTICE

A non-technical skills training program was developed for workers in a railway yard. One of the key areas identified in the training needs analysis was the non-technical skill of effective task management. The following is an excerpt from the competency specification, which also sets out the performance criteria that were used as the basis for assessment during an on-the-job training day.

TASK MANAGEMENT WITHIN THE RAILWAY YARD

This unit involves the core skills of managing multiple tasks in the environment of shunting operations in the yard. This unit is relevant to drivers, rail operators and panel operators.

ELEMENT		PERFORMANCE CRITERIA
1. Plan tasks	1.1.	All shunting activities are planned in consultation with the work team to ensure the safest and most efficient sequence of activities.
	1.2.	Specific tasks are assigned to individuals, and all team members are aware of each other's roles and responsibilities.
	1.3.	Plans, and any changes to plans, are communicated to all personnel working in the yard to ensure that situation awareness is maintained.

Non-technical skills have not always been a focus in competency-based training, and technical performance is often more easily described both in a competency specification and in the performance criteria that are used for assessment. For instance, the task of maintaining constant altitude in an aircraft is easily described in terms of the proper use of aileron and trim, and the performance criterion for assessment is easily described in terms of maintaining desired altitude within plus or minus 50 feet. However, as described in the 'Task Management within the Railway Yard' example, it is still possible to develop competency specifications that set out performance criteria for assessment of non-technical skills.

PRINCIPLE TWO: ASSESSING NON-TECHNICAL SKILLS SHOULD ADOPT A COMPETENCY-BASED APPROACH

Competency-based training (CBT) supports a structured approach for assessment of non-technical skills that focusses on assessment of integrated work performance, in authentic settings, and against specified criteria for competent performance.

Assessing Skills in Context: Behavioural Marker Systems

The most common, and highly effective, method for assessing non-technical skills is through the use of a behavioural marker system. A behavioural marker system can be defined as a framework that sets out observable, non-technical behaviours that contribute to superior or sub-standard performance within a work environment.[1] The first behavioural marker systems were developed to support formal assessment of the outcomes of crew resource management training in aviation. Today, behavioural marker systems form the foundation of the assessment of non-technical skills in an extremely diverse array of high-risk work environments.

Key Characteristics of Behavioural Marker Systems

Although used in diverse industries, behavioural marker systems share a set of common characteristics. These are as follows:

Observable Behaviours: The first characteristic of a behavioural marker system for non-technical skills is that they are constructed to identify and rate actual observable actions of individuals and teams. They do not focus on attitudes or personality traits, but rather, pivot around aspects of performance that can be objectively observed during normal operations or in training activities. They use word-pictures to describe examples of optimal and sub-optimal performance for a given non-technical skill in context.

Related to Safe and Efficient Operations: Non-technical skills are defined as skills that contribute, positively or negatively, to safe and efficient operations in high-risk work environments. Accordingly, the focus of any behavioural marker system is primarily on behaviours that have been established as critical to safe and efficient operations.

Taxonomic Structure: Most behavioural marker systems are structured as a taxonomy of skills and exemplar behaviours.[2] From a top-level broad domain such as communication, several constituent skills are defined, such as assertiveness, communication environment, and sharing information. Performance in each of these constituent skills is then defined using a set of behaviours that describe both good and poor performance. Figure 5.1 illustrates the generic hierarchical structure used in non-technical skills behavioural marker systems such as anaesthetists' non-technical skills (ANTS)[2] and Non-Technical Skills for Surgeons (NOTSS),[3] where the term *category* is used to describe a domain of non-technical skill, such

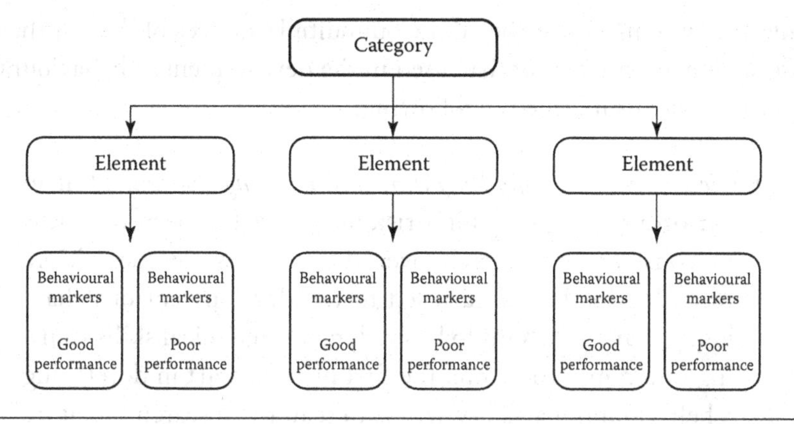

Figure 5.1 Generic structure of a behavioural marker system.

as communication, and *element* is used to describe a specific skill, such as assertiveness.

Cognitive and Social Skills: Most behavioural marker systems describe a set of both cognitive and social skills. While cognitive skills, such as maintaining situation awareness, are sometimes difficult to observe, the behavioural marker system must describe observable actions that can be shown to be evidence of the underlying cognitive skill. For instance, information acquisition strategies and a trainee's descriptions of system state are examples of specific behaviours relating to the underlying cognitive skill.

Provide a Common Vocabulary: One of the most important aspects of behavioural marker systems is that they provide a common vocabulary for an organisation to use with respect to non-technical skills.[1] By naming areas of skill that are deemed critical for performance, it is possible to achieve alignment between the training syllabus, competency assessment and normal everyday operations. Moreover, the behavioural marker system effectively sets out standards of expected performance by describing exemplar behaviours.

Development of Behavioural Markers

Most of the commonly used behavioural marker systems have been developed using a specific evidence base and have been subjected to validity and reliability assessments. Common to most behavioural marker systems is that they draw on multiple sources of data in their development. Sources of data used in the development of behavioural marker systems include the following:

Review of the Scientific Literature and Existing Systems: Human performance in most high-risk industries has been the focus of considerable research over the last few decades. More recently, this has in turn led to the development of behavioural marker systems to be used in non-technical skills training programs. Therefore, the first place to start in developing a behavioural marker system is to scan the research literature (Google Scholar is a useful tool) and also undertake a quick review of industry best practice.

Incident and Accident Reports: The first source of data relates to the output from the investigation of instances where safety has not been maintained. Incident and accident reports, if based on high-quality investigation processes, should identify the underlying human factors issues and areas of non-technical skill that were implicated as part of the causal chain of the safety occurrence. As described in the previous chapter, in most organisations, there should be an explicit link between non-technical skills training programs and the organisation's safety management system. While analysis of incidents and accidents by an organisation can inform training needs analysis, from the perspective of assessing non-technical skills, data from within the safety management system can also be used to identify specific behavioural markers of good and poor performance.

Cognitive Task Analysis: Developed from traditional task analysis, which sets out to deconstruct work performance into individual tasks and constituent actions, cognitive task analysis (CTA) is a method used to describe the mental processes that support work performance alongside the behavioural elements.[4] The processes for undertaking cognitive task analysis include methods such as cognitive interview techniques, verbal report methods in which expert operators describe what they are doing/thinking, and process mapping information, communication and decision pathways.

Interviews/Focus Groups with Expert Operators: Experienced operators are an excellent source of information relating to the non-technical skills giving rise to safe and unsafe situations. One common method for eliciting this information is through the use of a critical incident interview technique whereby experts are asked to recall and describe their management of a critical incident.[5,6] The interviewer then explores with the expert the specific skills they think are important for managing such events and elicits examples of good and poor performance.[2,7]

Once data have been collected using one or more of these tools, a competency specification can then be built for the relevant non-technical skills as described in the previous section.

EXAMPLE IN PRACTICE

A utilities company had several incidents within their control room operations that impacted significantly on service delivery. The incident investigations revealed that breakdowns in coordination between control room and field staff were an issue. The company engaged the services of a human factors specialist to assist them with developing a training program specifically for the non-technical skills associated with task coordination.

The human factors specialist used three main methods in gathering data to identify training needs and develop a behavioural marker system. First, all relevant incident investigation reports were reviewed, and the specific breakdowns in coordination were analysed. Second, the human factors specialist undertook observations in the control room and in the field, asking operators to verbalise the processes associated with coordinating activity between the control room and the field. Finally, a draft training syllabus and curriculum including a competency specification and behavioural markers were subjected to review and refinement in a workshop of subject matter experts and key stakeholders.

The following is an excerpt of the competency specification and behavioural markers used in the training program.

TASK COORDINATION – CONTROL ROOM AND FIELD OPERATIONS

This unit will enhance the core skills of managing multiple tasks in the environment of shunting operations in the yard. This unit is to be undertaken by small groups comprising both control room and field operators during a simulated valve failure.

ELEMENT	PERFORMANCE CRITERIA
3. Coordinated Failure Diagnostics	3.1 Both the field operator and control room operator will verbally describe the relevant indications in front of them;
	3.2 The control room operator will synthesise this information, and generate a provisional diagnosis;
	3.3 The control room operator will share the provisional diagnosis with the field operator;
	3.4 The field operator will identify and communicate any alternative diagnoses; and
	3.5 The control room operator and field operator shall agree on a diagnosis or gather further information if any uncertainty exists.

BEHAVIOURAL MARKERS OF GOOD PRACTICE

- Standard phraseology is used to describe objects, failures and processes;
- Read-back is used to confirm information is received correctly;
- All relevant information is communicated accurately;
- Alternative diagnoses are generated and shared between the control room operator and the field operator; and
- Any uncertainty is communicated and subjected to further analysis towards resolution.

BEHAVIOURAL MARKERS OF POOR PRACTICE

- Incorrect or non-standard phraseology is used;
- No read-back is provided to confirm information is received correctly;
- Potentially important pieces of information are not communicated;
- Alternative diagnoses are not generated and discussed; and
- Uncertainty is left unresolved.

Validity, Reliability and Sensitivity

Once a set of behavioural markers have been developed, it is critical that they are subjected to an assessment of their validity and reliability before being used in the process of training and assessment.

Validity refers to the degree to which the behavioural marker system actually assesses what it claims to be assessing. If the developmental process described in this chapter has been followed, the behavioural marker system will already demonstrate a high degree of validity. There are a number of different elements of validity,[9] as described in Table 5.2.

Table 5.2 Aspects of Assessment Validity

VALIDITY CONSIDERATION	DESCRIPTION
Content validity	Definition: The degree to which the assessment tool tests the domain of skills as used in practice, such as assessing the different levels of situation awareness. Establishing: Compare content of assessment against competency specification and identify any missing or irrelevant elements of the assessment.
Construct validity	Definition: The degree to which the assessment actually measures the non-technical skill it suggests it is assessing, such as task management and not decision-making. Establishing: Compare performance on assessment against another measure of performance on the same task.
Criterion validity	Definition: The degree to which performance on the assessment compares with other measures of actual task performance, such as performance in the simulator compared with the operating theatre. Establishing: Compare whether people who score low or high on the assessment perform well or sub-optimally on real work.
Consequential validity	Definition: How well the assessment achieves positive outcomes, such as highlighting specific areas for improvement, and avoids negative outcomes, such as failing trainees who are actually competent. Establishing: Through the use of evaluation measures such as feedback and monitoring performance.

Sources: Isaac, S. and Michael, W.B., *Handbook in Research and Evaluation for Education and the Behavioral Sciences*, San Diego, CA, EdITS, 1995; and Linn, R.L. and Gronlund, N.E., *Measurement and Assessment in Teaching*, Upper Saddle River, NJ, Prentice-Hall, 2000.

Reliability, on the other hand, refers to the consistency of assessment results.[8] It can be considered as the next most important factor to validity and is critical to the assessment of non-technical skills.

As with validity, there are multiple forms of reliability, and each can be established through different techniques as described in Table 5.3.

Sensitivity is the third major consideration in ensuring effective assessment of non-technical skills. Sensitivity refers to the degree to which the behavioural marker system is able to detect good and poor performance in a specific area of non-technical skill: in other words, how well the system enables the user to analyse and rate performance. An assessment system that lacks sensitivity is one that is unable to help the user in differentiating nuanced aspects of performance. In general terms, the less generic a system is, the more sensitive it is likely to be.

With respect to behavioural marker systems, sensitivity is measured and evaluated in several different ways. One of these is to use

Table 5.3 Aspects of Assessment Reliability

RELIABILITY CONSIDERATION	DESCRIPTION
Stability over time	Definition: The degree to which the assessment yields the same results at different points in time, such as assessing situation awareness every six months. Establishing: Using the test-retest method whereby the same assessment is given at different time intervals.
Between assessment forms	Definition: The degree to which the assessment yields the same results when given in different forms, such as assessing communication skills during different simulation scenarios. Establishing: Comparing the assessment results of individuals when undertaking different forms of the assessment.
Between assessors	Definition: The degree of consistency in ratings between different assessors. Establishing: Measuring inter-rater reliability during instructor training.

Source: Linn, R.L. and Gronlund, N.E., *Measurement and Assessment in Teaching*, Upper Saddle River, NJ, Prentice-Hall, 2000.

a standard set of video-recorded scenarios in which both good and poor performance are captured. Expert raters (typically those involved in scripting the scenarios and the performance on the trial videos) then use the behavioural marker system to rate performance. Then, typical users are trained in the use of the behavioural marker system, and their ratings are compared with the reference expert ratings. The degree of similarity between the two ratings can then be used as a measure of sensitivity. The smaller the difference between typical users and expert ratings, the higher the sensitivity of the tool.[3]

Rating Scales and Scoring

While a set of behavioural markers is used to define and describe specific aspects of each domain of non-technical skills, actual assessments are usually made using some form of rating scale built into the system. Performance ratings are the most common form of assessment tool used in many settings and have a long history of use in the measurement of job performance.[10] Performance rating in its most basic form involves a rater evaluating an individual's or team's performance against some pre-determined criteria. It does not rely solely on objective metrics, but rather, involves a degree of judgement on the part of the rater.

Table 5.4 Example Rating Scale for Non-Technical Skills

1 = POOR	2 = MARGINAL	3 = GOOD	4 = OUTSTANDING
Observed performance had safety implications	Observed performance was barely adequate	Observed performance was effective	Observed performance was truly noteworthy

Source: Klampfer, B. et al., *Enhancing Performance in High-Risk Environments: Recommendations for the use of Behavioural Markers*, Ladenburg, Daimler-Benz Stiftung, 2001.

The simplest rating scales use a binary 'pass/fail', 'competent/not yet competent' or 'acceptable/unacceptable' to assess performance. By limiting the degree of choice to a simple binary, it is thought that greater levels of consistency between assessments can be achieved.

However, frequently, a higher level of granularity is desired to differentiate performance and guide the debriefing process. To this end, four- or five-point scales are often used in the assessment of behavioural markers of non-technical skill. The rating scale from the University of Texas' non-technical skills rating system as used in the Line Operations Safety Audit (LOSA) is illustrated in Table 5.4.

The four-point rating scale developed by the University of Texas sets a standard score of 3, where it is assumed that the performance is that of a competent operator. A score of 4 is used to demonstrate exemplary performance. At the other end of the scale, a score of 1 is used when the performance against the non-technical marker is directly linked to a negative safety outcome. A score of 2 is then used to demonstrate unsatisfactory performance but without safety being directly compromised on this occasion.

A variety of other rating scales have been developed and can be found in the various non-technical skills assessment tools such as NOTECHS,[11] NOTSS[3] and ANTS.[2]

PRINCIPLE THREE: THE USE OF A BEHAVIOURAL MARKER SYSTEM IS AN EFFECTIVE, AND WIDELY USED, TECHNIQUE FOR ASSESSING NON-TECHNICAL SKILLS

Behavioural marker systems are highly effective in the assessment of non-technical skills. They can be developed using a cross-section of data sources and are used to describe optimal and sub-optimal performance in actual work tasks.

Generic versus Specific Behavioural Marker Systems

The majority of behavioural marker systems are very generic in nature. That is to say, they describe a set of overarching non-technical skills that have been demonstrated to relate to safe and efficient performance across a wide range of situations. For instance, a communication skill such as *assertiveness* can be described in terms of advocating a position without being either passively accepting of something you don't agree with or overly forceful or aggressive. Described in this way, this skill is quite generic and easily transferred across a wide range of situations.

The use of a generic system has a range of advantages. First, it is *portable*, inasmuch as it can be used across a variety of different training settings and across many different workplace scenarios. Second, it creates *efficiencies*, inasmuch as it only needs to be created and validated once. Further, those who use it also need only to be trained and calibrated to a single generic system. However, the use of a generic system brings with it limitations with respect to specificity and sensitivity.

The use of markers that describe only generic aspects of non-technical skills may not provide instructors/assessors with sufficient detail to understand the nuances of performance in specific scenarios. Therefore, to describe the behaviours that demonstrate optimal performance in a non-technical skill as they relate to a specific scenario, a different approach is required.

In contrast to a generic behavioural marker system, it is possible to create a much more tailored set of behavioural markers that are specific to a discrete training scenario. Take, for example, the non-technical skill category of *situation awareness*, and its first element, *information gathering*. As illustrated in Table 5.5, a typical generic set of behavioural markers sets out specific actions at a very high level. However, by specifying a set of behaviours that relates specifically to a given scenario, a much more powerful assessment tool can be created.

Creating a scenario-specific set of behavioural markers has a number of significant benefits. First, they are much more *sensitive* to being able to identify and assess performance, as the behaviours expected of

Table 5.5 Generic versus Specific Behavioural Markers for Situation Awareness – Information Gathering

GENERIC	SPECIFIC
SCENARIO: GLOBAL	SCENARIO: LOSS OF POWER AT HOME
Frequently scans the environment	Identifies loss of power to an appliance
Monitors actions of others	Checks appliance is properly plugged into socket
Actively seeks missing information	Identifies other appliances are affected
Maintains vigilance	Verifies status of circuits in fuse box
	Identifies tripped circuit

good (and poor) performance are very clearly laid out for the instructor/assessor. In turn, this would logically result in an increase in *inter-rater reliability*. No longer are the instructors/assessors relying on their own judgement in interpreting how a generic set of behaviours might look with respect to a specific scenario. Rather, they have at their fingertips a specific set of behaviours identified just for that single scenario.

In the domain of medical education, considerable research has explored the relative benefits of global scales, such as a generic behavioural marker system, in contrast to checklists, which specify specific actions that are deemed to be critical to achieving the task goals.

A recent review of this research highlighted that global rating scales have higher degrees of reliability, can be used across multiple tasks, and may be better at capturing nuanced elements of expertise than checklists that produce scores on the presence or absence of a specific action.[12] Global rating scales allow more judgement on behalf of the assessor but perhaps lack specific guidance on identifying the actual behaviours that demonstrated good or poor performance. Judgements can be more generalised but can lack specificity.

However, if using a highly specific checklist of behavioural markers, attention needs to be paid in the debriefing of training to ways in which the specific example behaviours can be generalised to other situations in the work environment. If this is not done, there may be inadvertent issues created with respect to limitations of transfer of training, as discussed in the previous chapter.

PRINCIPLE FOUR: A BALANCE NEEDS TO BE ACHIEVED BETWEEN GENERIC SETS OF BEHAVIOURAL MARKERS AND CHECKLISTS OF DESIRABLE/UNDESIRABLE ACTIONS IN A GIVEN SCENARIO WHEN ASSESSING NON-TECHNICAL SKILLS

While generic sets of behavioural markers may be easily deployed across multiple training scenarios and allow more nuanced judgement of expertise, specific checklists of actions provide good guidance to the instructor as to the desirable actions to be performed in a specific scenario. Deploying a global rating scale of generic markers in assessment, but providing instructors with good guidance on the specific behaviours that are expected in the specific training scenario, is likely to achieve the balance between generic and specific assessment tools.

Integration with Technical Assessment

The distinction between technical and non-technical skills is really a false divide. Effective performance in high-risk industries is better defined as being a product of a set of competencies, where the technical and non-technical aspects are intertwined. While we still differentiate non-technical skills for reasons of simple explanation and convenience, eventually we will need to move to more integrated forms of training and assessment.

Already, there exist some examples where the assessment of technical and non-technical skills have been very successfully integrated. Once such system was developed in the aviation domain, as an element of the Proficiency Assessment and Training System (PATS), developed by Captain Simon Henderson then at Ansett Australia. This approach first seeks to integrate technical and non-technical aspects of performance by first identifying a positive or negative performance outcome, such as landing the aeroplane outside the touchdown zone. The instructor/assessor then sets out to identify the *reasons* why that poor outcome occurred. A checklist approach is used, including domains of both technical and non-technical skills.[13] Table 5.6 provides an overview of these 'reason codes'.

Table 5.6 Technical and Non-Technical Reason Codes Used in PATS

TECHNICAL	NON-TECHNICAL
M: Manipulation	C: Communication
K: Knowledge of systems and procedures	W: Workload management
A: Automated systems comprehension and use	S: Situational awareness
E: Execution of procedures	P: Problem-solving and decision-making

Source: Phelan, P., *Flight Global*, 1999.

The approach embedded in PATS is beneficial for several reasons. First, it adopts from the outset a mindset that safe performance outcomes are the result of both technical and non-technical skills. Second, it embodies a training philosophy that assessment enables subsequent training interventions to be targeted at areas of need. To this end, assessment identifies areas of both technical and non-technical deficits that can then be worked on by the trainee.

PRINCIPLE FIVE: AN INTEGRATED APPROACH TO ASSESSING NON-TECHNICAL SKILLS, IN WHICH TECHNICAL ASPECTS OF PERFORMANCE ARE ALSO EVALUATED, IS HIGHLY DESIRABLE

In all high-risk industries, safe and efficient performance is the product of both technical and non-technical skills. Assessing one of these without reference to the other does not adequately reflect an operator's complete competency in task performance. Therefore, an integrated approach to assessing technical and non-technical skills is desirable.

Jeopardy Assessment: Pass or Fail?

One of the areas of significant debate with respect to the assessment of non-technical skills relates to whether an individual can be assessed as 'failing' a non-technical skills training program, which we will refer to here as *jeopardy* assessment.

As discussed in Chapter 1, it is an underlying assumption of this book that non-technical skills form essential elements of competency, and without formal training and assessment of non-technical skills, an operator cannot be deemed to have all the skills required to perform a safety-related role in a high-risk industry. However, this view is not held universally.

Objective Assessment

One of the original reasons for resistance to jeopardy assessment of non-technical skills relates to the initial lack of formally validated objective assessment techniques. It is indeed more difficult to rate behaviours, particularly when they are the result of 'hidden' cognitive processes, than it is to rate technical performance, which often has an associated objective measurement such as a patient's blood pressure or the train's speed. However, as we have already outlined in this chapter, high-risk industries across the globe have developed valid and reliable assessment frameworks. If the processes outlined in previous sections for the design and validation of assessment frameworks were followed during their development, the argument that assessment is highly subjective is no longer convincing.

Individual or Team Performance?

Another of the criticisms of jeopardy assessment of non-technical skills is that they are often deployed in a team-based work environment. Therefore, performance is a product of the team and not the individual, so how can individual assessments of competency be made? While this argument may ring true if a purely outcome-based approach to assessment is adopted, focussing on outcome at the expense of examining an individual's non-technical knowledge and skills is inappropriate. As will be discussed in following sections, focussing on the outcome of performance is actually likely to bias assessment of the underlying skills. The benefit of a well-designed non-technical behavioural marker rating system is that it can be deployed at the individual level to identify the specific behaviours that gave rise to the positive or negative technical outcome.[11] For instance, it could be the poor situation awareness of the captain and the poor assertiveness of the first officer that both contributed to the poor outcome.

Linked to Technical Failure?

The widely used European NOTECHS system for rating flight crew non-technical performance adds another dimension to the discussion around whether a pass/fail judgement can be made with respect to non-technical skills. One of the explicit assumptions of NOTECHS is that any rating of poor performance in a non-technical domain can only be made if there is an actual demonstrable link to a technical consequence. If there is no actual (or potential) impact on the technical outcome of safe flight associated with the poor non-technical performance, the pilot's performance cannot be rated as unacceptable.[11] There are two possible explanations for this constraint on assessment. The first is that it is a compromise in deference to the vocal old school, who prioritise technical performance over non-technical performance. The second is an attempt to reflect the need to consider technical and non-technical performance as an integrated whole.

Is Poor Non-Technical Performance Reason Enough to Fail?

From my personal perspective, consistently demonstrating poor non-technical performance is sufficient to be deemed overall lacking in competence and requiring remedial training. We now have the tools for valid and reliable assessment of non-technical skills. Further, as was elaborated on when describing the assumptions of this book, non-technical skills have been established to be directly related to poor safety outcomes. Accordingly, for someone to be deemed competent in a safety-critical role in any high-risk industry, they must be competent in all relevant domains of non-technical skill.

PRINCIPLE SIX: AS A CRITICAL DETERMINANT OF SAFETY IN HIGH-RISK INDUSTRIES, POOR NON-TECHNICAL PERFORMANCE SHOULD BE REASON ENOUGH TO BE DEEMED NOT COMPETENT

Someone who lacks critical technical knowledge and skills in the operation of equipment such as the flight management computer in an aeroplane, or the anaesthetic machine in the operating theatre, would not be deemed competent for unsupervised work in those domains. Similarly, someone who lacks critical knowledge or skills relating to task management or communication should also be seen as lacking essential competencies in any high-risk industry.

Training Instructors, Calibration and Inter-Rater Reliability

Training the Non-Technical Skills Instructor and Assessor

Much has been written elsewhere with respect to the training required for being an effective instructor and assessor in competency-based training environments, and all of this translates directly to non-technical skills training programs. In general, the core skills that need to be developed by the instructor and assessor include the basic principles of how adults learn and effective facilitation techniques for the classroom, simulator and on-the-job forms of non-technical skills training.[14]

Worldwide, there are many vocational training programs that can be taken to develop effective instructional techniques, and these should all provide a solid foundation for non-technical skills trainers and assessors. Over and above these general training programs for instructional skills, the specific knowledge and skill requirements for non-technical skills instructors and assessors are highly dependent on the training context. However, a number of general principles can guide the appropriate selection of non-technical skills instructors and assessors.

First, the requirement for technical expertise in the work domain is important. As has been discussed previously in this chapter, the very best non-technical skills training programs integrate the training of both technical and non-technical competencies. However, there are also many exceptions to this rule of thumb, especially when a team with both technical and non-technical expertise can undertake instruction and assessment. In many work environments, a Human Factors specialist may work with a technical expert and provide enhanced levels of non-technical instruction and assessment.

The critical requirement is that the instructor and assessor has extensive knowledge of the non-technical skills being trained, has a detailed understanding of how these skills contribute to enhanced or sub-optimal performance in real-world operations, and has at least a good enough working knowledge of the technical operation to be able to contextualise the non-technical skills being trained in the context of normal and non-normal operations. Similarly, acceptance by trainees that the trainer has valid expertise and sufficient knowledge and skill is extremely important. In some domains, such as medicine and aviation, the culture is such that non-operational staff and junior staff might not be seen as sufficiently skilled to undertake non-technical skill training and assessment. Beyond an acceptable level of knowledge and skill and an appropriate level of *gravitas*, non-technical skills instructors and assessors need to have their competencies developed in terms of training, assessment and facilitation techniques.

Inter-Rater Reliability

Previously in this chapter, we introduced the notion that an assessment system needs to demonstrate *reliability*, which refers to the accuracy and consistency of the assessment tool. With respect to the assessment of non-technical skills in high-risk industries, the concept of *inter-rater reliability* is perhaps the most important feature of reliability, given that often a large number of instructors will be responsible for assessment across the workforce.

In an ideal world, the same performance in a scenario would be rated in exactly the same way by the same instructor on multiple occasions, or indeed by all the different instructors in the training department. In reality, there are a number of barriers to achieving perfect inter-rater reliability, and the research literature highlights that often only modest inter-rater reliability is achieved when using some of the most common behavioural marker assessment tools. Much of the modest level of inter-rater reliability may be due to sources of rating error, such as biases, which will be examined shortly. However, the lack of high levels of inter-rater reliability probably reflects the brevity of training instructors and assessors and spending time on the calibration of ratings.

EXAMPLE IN PRACTICE

A non-technical skills training program was being developed for intensive care teams at a large hospital to enhance skills in situation awareness and decision-making. A training needs analysis had been completed, and competency specifications with associated behavioural markers had been written. The training curriculum had also been finalised, and this involved core knowledge development in a workshop followed by a full team simulation with the scenario of a patient being admitted to the unit with significant post-operative complications and a post-simulation debrief.

All that remained was to train the instructors in assessing the non-technical skills described by the training needs analysis and competency specifications with associated behavioural markers. A train-the-trainer day was arranged, and videos of examples of good, average and poor performance in the scenario had been recorded.

To facilitate calibration of the instructors, a simple inter-rater reliability tool was used, based on the within-group agreement (r_{wg}) statistic. The instructors viewed each video, and afterwards, one of the non-technical skills was assessed, and the instructors made a rating on a four-point scale. These scores were collated and input into a spreadsheet that made the calculation of the r_{wg} statistic. A value greater than .7 was deemed to represent sufficient level of agreement. However, all instances where there was disagreement on scoring were discussed, and a calibrated score for the performance on the video was agreed on as part of the instructor training and calibration process. The inter-rater agreement tool is shown here.

Inter-rater agreement (r_{wg})
$r_{wg} = 1-(SX^2/\sigma E^2)$

VIDEO		**Post operative complications**		
Non-techncial skill		**Information gathering**		
Instructor	**Rating**		**VAR**	0.267857143
Instructor 1	1		**POP VAR σE²**	1.25
Instructor 2	2		**r_{wg}**	**0.79**
Instructor 3	1		Mean	1.375
Instructor 4	1		**Calibrated**	**1**
Instructor 5	2			
Instructor 6	2			
Instructor 7	1			
Instructor 8	1			

The degree of inter-rater reliability achieved in the assessment of non-technical skills is a product of the assessment tool itself combined with the knowledge and skills of those undertaking the assessment. The reliability of the assessment tool has been discussed in detail earlier in this chapter. Just as with non-technical skills themselves, developing skills in the assessment of these skills takes considerable knowledge development and practice in the application of a behavioural marker system.

PRINCIPLE SEVEN: INSTRUCTOR TRAINING, CALIBRATION AND INTER-RATER RELIABILITY ARE CRITICAL ASPECTS OF THE EFFECTIVE ASSESSMENT OF NON-TECHNICAL SKILLS

Instructors and assessors in non-technical skills training programs require general skills in instructional techniques, specialised knowledge in non-technical skills domains and calibration in assessment techniques to ensure inter-rater reliability.

Biases and Assessment Error

There are a number of well-known biases that can impact on the reliability of rating the non-technical performance of individuals and teams. The first group of biases that can negatively impact on rating non-technical skills all relate to situations in which one aspect of performance becomes the main focus of assessment, and in turn, that aspect of performance then influences other aspects of assessment. A second set of biases relate to general patterns in the way in which an assessor makes their ratings. The final set of biases relate to the influence of normative assessment instead of the criterion-referenced assessment at the heart of a competency-based training approach to non-technical skills development.

Halo Effect

Perhaps the most common of the biases that can influence raters' scores is the 'halo' effect, in which a general conception of the individual or team performance is made, and that subsequently influences all ratings. This is especially the case when a particularly good, or poor, outcome is achieved, and this outcome 'masks' the areas of sub-optimal performance.

EXAMPLE FROM PRACTICE: HALO EFFECT

During a simulator exercise, a control room team are able to quickly resolve a complex malfunction impacting on a large part of the electricity network. It was one of the quickest restorations of power the instructor had seen, and the impact on service delivery to customers was minimal. The instructor therefore gives good scores across all the categories, even though the coordination of the team, the evaluation of risk and the identification of alternative solutions were lacking. These failures in non-technical skills led to a solution that while timely, actually compromised safety. However, due to the positive outcome, consistently high scores were given by the instructor.

The halo effect is best detected by examining the amount of variability between the individual categories rated by an individual. If a low degree of variance, or a high degree of correlation, is seen consistently between ratings on all the different categories assessed, the ratings might be influenced by the 'halo' effect.[15] This is especially the case if ratings are consistently high. Further training in the use of the assessment and rating tool is the most effective mechanism to address instances of the halo effect. Similarly, the presence of the halo effect may indicate deficiencies in the assessment tool, such that the sub-elements are inadequately specified, and raters can't discriminate between areas of good and poor performance.

Visceral Bias

Another form of rating bias is called, quite symbolically, *visceral bias*, whereby a deep intuitive feeling influences ratings of actual performance. This describes the situation when an assessor uses their 'gut feel' rather than an objective assessment of performance to derive their rating. This bias was originally used with respect to describing the bias associated with the way in which first impressions of a patient can influence clinical diagnostic decision-making.[16]

EXAMPLE FROM PRACTICE: VISCERAL BIAS

A check and training captain is scheduled for a check ride with a first officer who is well known for making highly inappropriate comments towards cabin crew. The check and training captain just does not like this first officer at all, and as a result, he rates the performance of the first officer as lower than it actually was in practice.

As visceral bias describes how an assessor develops both positive and negative feelings with respect to individual trainees or teams, the influence of this bias is difficult to detect externally. It has been suggested that the only effective way to mitigate this bias is to consciously reflect on the potential that one's emotions might be influencing judgement and to actively make efforts to ensure that judgements are objective, rather than affective, in nature.[17]

Central Tendency

The bias of central tendency relates to the natural propensity to award scores around the middle of the range and avoid scoring performance as extremely good or extremely poor. This has been seen to be an unwanted feature of many work-based assessment systems in high-risk industries over the years.

EXAMPLE FROM PRACTICE: CENTRAL TENDENCY

Jane is a new simulator instructor on the surgical non-technical skills training program at a large downtown hospital. The scenario involves the surgeon having to assist in resolving an argument between the anaesthetist and the scrub nurse to keep the team task-focussed. Given the fact that she is making ratings of the performance of senior colleagues, she finds it difficult to score performance as poor. Likewise, she does not want to be seen to be too soft, so she plays it safe by never using the extremes of the rating scale.

Central tendency bias is easily detected by looking for a distribution of scores that are close to the mean or median score on the scale and exhibit a small standard deviation. This is called *leptokurtosis*. Instead of the normal distribution of scores looking like a rounded bell shape, the distribution looks more like a spike in the middle. As central tendency is often driven by cultural factors in high-risk industries, it is best addressed by empowering raters to highlight the extremes of performance and giving them the debriefing tools to ensure that even elements of poor performance can be turned into positive learning outcomes.

Leniency and Severity

Another set of biases that can impact on the reliability of the assessment of non-technical skills are individual in nature. Over time, an individual can develop specific patterns of rating that might deviate from the norm. The first of these is *leniency error*, whereby a rater is inclined to give higher ratings than others. Similarly, *severity error* is when a rater is predisposed to focus on the negative aspects of performance and give consistently lower scores.

EXAMPLE FROM PRACTICE: LENIENCY AND SEVERITY

John is a tutor driver who has been in the rail industry for over 40 years. He is now a senior instructor whose role it is to certify trainee drivers as competent to operate without supervision. He has seen a lot in his time, and he wants to make sure that safety will not be compromised due to allowing a driver to operate unsupervised when they are not yet ready. He therefore likes to focus on any aspect of sub-optimal performance. He prides himself on being a hard marker.

This form of bias is relatively easy to detect over time when individual rating patterns, and especially mean scores, can be compared among a group of raters in an organisation. The most effective way to address a bias towards leniency is to provide visibility of an individual rater's scoring patterns in contrast to those of their peers. A number of instructional management systems support this process, providing automated analysis and reporting of the variation between instructors' patterns of scoring.

Contrast Effect

The *contrast effect* refers to expert raters comparing the performance of trainees with their own performance or the performance of other trainees. This can be described as a reversion to normative assessment (comparing and ranking performance between peers) rather than the criterion-referenced assessment that is critical to competency-based training.[18]

EXAMPLE FROM PRACTICE: CONTRAST EFFECT

Three anaesthetic registrars attended a simulation-based training workshop that focussed on team management during anaesthetic crises. Each trainee took turns in playing the role of the anaesthetist during a simulated patient crisis in the operating theatre. The first anaesthetist displayed outstanding performance in managing the situation. The second and third anaesthetists displayed competent management; however, they received lower ratings of their performance. Rather than assessing solely on the basis of the criteria set out in the behavioural marker system, the assessor compared their performance with that of the first trainee. The second and third anaesthetists were unduly rated as poor performers, when in reality, their performance was competent.

Contrast effects are difficult to overcome and are resilient to the use of multiple assessors, as the same order effects will influence each assessor. They are best detected by random audit of assessments made by instructors that do not include a series of sequential assessments.

A summary of each of these biases and the techniques used to avoid or overcome them is provided in Table 5.7.

Table 5.7 Summary of Biases Impacting on Performance Assessment and Their Remedies

BIAS	REMEDY
Halo effect: A general assessment of performance influences ratings of all aspects of performance.	Further training in the assessment tool to improve discrimination between each of the domains and specific constituent skills.
Visceral bias: A rater's initial emotional response to the individual or team influences the rating of their performance.	Prompting raters to make explicit any emotional response to the individual or team and to critically reflect on whether these might influence rating. Providing assessment tools that demand reference to observed actions in forming a rating.
Central tendency: The unwillingness to use the extremes of a rating scale.	Empowering raters to identify aspects of poor and excellent performance and providing tools to ensure that the extreme ratings can be used to create positive learning outcomes.
Leniency and severity: An overall tendency to give ratings either above or below the average score.	Develop a continuous monitoring program of individual raters' assessments compared with those of their peers. Provide regular opportunities for calibration of raters within an organisation.
Contrast error: A tendency to make ratings on the basis of a comparison between trainees rather than independently and solely in relation to the assessment criterion.	Reinforce strict adherence to criterion-referenced assessment. Provide high-quality behavioural markers and checklists to support assessment decisions.

PRINCIPLE EIGHT: A SET OF BIASES THAT CAN NEGATIVELY IMPACT ON THE ASSESSMENT OF NON-TECHNICAL SKILLS MUST BE IDENTIFIED AND RECTIFIED ON A REGULAR BASIS

Processes for identifying and rectifying the influence of biases on the assessment of non-technical skills are an important component of an effective non-technical skills training program. It cannot be assumed that once trained, instructors and assessors will maintain consistent performance. Automated monitoring of assessments, as well as recalibration of instructors and assessors at regular intervals, is important.

Expert versus Self-Assessment

The use of self-assessment in a number of high-risk industries has been a feature of many training programs, and indeed, self-assessment is a fundamental component of promoting reflective practice in on-going professional development.[19] The ability for individuals or teams to self-assess as part of their debriefing process has the potential to save costs associated with requiring an instructor-led debrief and opens up opportunities for debriefing outside the context of formal training programs.[20]

However, while there is some evidence that self-assessment can be effective in technical domains, this is not always the case with respect to non-technical skills. Indeed, there is considerable evidence to suggest that even the most highly qualified professionals lack insight into their non-technical performance.[21,22]

To some degree, this is a classic example of the Dunning–Kruger effect,[23] whereby people who are unskilled in an area lack the metacognitive ability to realise their own incompetence. In this case, the trainees are both *unskilled* and *unaware* of it. The resolution to this problem of lack of insight is to embark on further skill development, as competency comes with the ability to better differentiate the quality of performance.

In short, self-assessment does play a role in on-going skill development[24] and should be used once basic skills have developed such that a trainee has sufficient insight to accurately identify good and poor aspects of their performance.

PRINCIPLE NINE: SELF-ASSESSMENT OF NON-TECHNICAL SKILLS SHOULD BE APPROACHED WITH CAUTION, AS TRAINEES OFTEN LACK INSIGHT INTO THEIR OWN PERFORMANCE

Unlike technical skills, there is considerable evidence that early in training, individuals lack insight into their non-technical performance. Accordingly, self-assessment should only be used once sufficient skills have been developed that a trainee can have sufficient insight into their own non-technical performance.

Debriefing Non-Technical Skill Performance

The process of post-event debriefing is important for non-technical skill development, whether it be after participating in role-play, simulation scenario or workplace-based training. Debriefing provides an opportunity for the learner to reflect on their own performance and/or that of their team and is critical to the consolidation of skills and knowledge as well as the identification of areas where things could be approached differently in the future. Recent studies have demonstrated the criticality of debriefing, to the degree that without debriefing, training participants show no improvement in their non-technical performance when subsequently assessed, compared with significant improvement in those who do participate in debriefing.[25]

During the 1990s, a significant body of work was undertaken by researchers at NASA with respect to facilitated debriefing of individual and team performance in the domain of non-technical skills. Researchers Key Dismukes, Lori McDonnell and Kimberly Jobe undertook a number of studies examining the techniques used by instructors in facilitating debrief of flight crew teams after line oriented flight training sessions.[26] These studies were undertaken at a number of airlines that had been early adopters, embedding crew resource management training into specific simulation-based training sessions. With the aim of describing evidence-based best practice, the group produced the definitive guide to facilitating debriefings.[27,28] They outline a set of key principles for best practice, which still provide excellent general guidance today and have also been reinforced and expanded on in other settings.[29]

Setting Goals and Expectations for the Debrief

First, the most effective debriefs begin with a short discussion of the goals and expectations for the discussion. This should outline each of the key areas discussed in the following sections and set the tone for critical reflection on team performance and the identification of strategies for enhanced performance. A suitable structure for a debrief involves contextualising non-technical skills relevant to the scenario, analysis and evaluation of the individual's or team's performance in each of the key facets of non-technical skills relevant to the scenario, identification of areas to enhance performance, and application of those to everyday operations.[30]

Facilitation, Not Instruction

Second, debriefs should not be instructor-led or simply involve one-way feedback on performance.[14] Rather, the instructor should facilitate. That is to say, the instructor guides the conversation, steering the trainees to areas where they can self-critique and critically reflect on their performance. This is not an easy task, and it was an area met with considerable resistance from flight crew who did not wish to critique their own performance. However, as introduced previously in this book, critical self-reflection is an important aspect of on-going professional practice.

One technique used to achieve this in practice is called the *reverse* debrief, in which the participants are asked to open the discussion themselves. Ideally, participants will debrief themselves; hence the term *reverse debrief*.[31]

However, the metacognitive skills associated with the reflection on individual and team performance are the very skills of the reflective practitioner, as introduced in the previous chapter. It is known that developing these skills in the context of non-technical skills training has some degree of transfer into everyday professional practice.[32] In a facilitated debrief, participants are guided to analyse performance beyond simply identifying what was good and bad, also reflecting on why performance was sub-optimal and identifying strategies to enhance future performance.

Promoting Reflection with Use of Video and Other Tools

As a tool that could potentially enhance the debriefing process, video recordings of the training session began to be used in early aviation and anaesthesia non-technical skills programs. Reviewing the video of the training session was quickly seen to be of significant benefit to effective debriefing, as the team were able to sit back and observe their performance from the perspective of a third person. Video-assisted debriefing has since become commonplace in non-technical skills training programs.

Although the scientific evidence to demonstrate that video review after training significantly improves transfer of training is equivocal,[33] there is sufficient evidence to suggest that video enhances the debriefing process. The benefits of video-based debriefing include: (1) video accurately portrays performance in an objective manner; (2) video enhances reflective reasoning in experiential learning; and (3) video is associated with positive participant reactions.[34]

In some industries, other tools are available to instructors and participants to enhance their debrief. For instance, in aviation, advanced simulators provide a recording of all critical aircraft parameters and can provide an animated replay of the flight path, flight instruments and aircraft configuration. While this is especially useful in terms of technical performance, such technology can also be very useful when debriefing non-technical skills domains such as *situation awareness* and *decision-making*.

Questions, Probes and Prompting Deeper Discussion

One of the most important aspects of facilitation involves asking the crew questions that prompt a deeper analysis of their performance and reflection on strategies to enhance performance in the future. Some of the key questions and probes are provided in Table 5.8.

Active Listening Techniques

The use of specific communication strategies is another critical element of effective debriefing. First, *active listening* is a very effective strategy to keep momentum in a trainee-led discussion. This technique involves periodic paraphrasing of what the trainee has stated as a means of demonstrating engagement in the discussion without actively driving the discussion. Table 5.9 provides a brief overview of active listening techniques useful in the debrief situation.[26,27]

Second, and perhaps somewhat counter-intuitively, *silence* can be a very effective mechanism in debriefing. The NASA studies found that silence of a few seconds was effective. This can be explained by way of the inherent desire to 'fill a gap' in a conversation. If the instructor does not fill the gap, it is more likely that a participant will do so, ensuring that the debrief discussion remains participant-led.

Linking Debrief to Rehearsal

One of the aspects that are done least well in many non-technical skills training programs is the purposeful linking of debriefing in subsequent rehearsal. The axiom of 'practice makes perfect' is well

Table 5.8 Questions and Probes for Facilitating Debrief

QUESTION/ PROBE	DESCRIPTION
Set the scene	The facilitator orients the participants to a particular point in the training sequence to focus the discussion on specific events or actions.
Ask for reaction	The facilitator asks the crew to self-reflect on a part of the training sequence and identify what went well and what didn't.
Lead the crew	The facilitator responds to a lull in the discussion by directing the crew to a specific topic for reflection, typically a relevant non-technical skill.
W words	The facilitator uses one of the 'what, where, when, how, why?' questions to prompt further analysis.
Deepen	The facilitator asks the participants for more detailed analysis of the situation and areas for enhanced performance.
Follow-up	The facilitator picks up on a topic brought up by a participant and asks for further analysis, demonstrating the participant-led approach.
Turn-back	The facilitator diverts any questions from participants back to them rather than providing their own opinion. This promotes deeper critical self-reflection.
Drawing out	The facilitator directs questions to participants who are not actively participating in the debriefing.
Redirect	The facilitator redirects questions of other participants to participants who are not actively participating in the debriefing.
Ask to expand	The facilitator asks a participant to expand on what they are saying or asks a quiet participant to expand on what someone else has suggested.
Reword	The facilitator simply rewords the question to facilitate further analysis and discussion or to restart a stalled discussion.

Sources: McDonnell, L.K., *Facilitation Techniques as Predictors of Crew Participation in LOFT Debriefings*. Ames Research Center Moffett Field, California: National Aeronautics and Space Administration, 1996; and McDonnell, L.K. et al. *Facilitating LOS Debriefings: A Training Manual*, Ames Research Center Moffett Field, California: National Aeronautics and Space Administration, 1997.

Table 5.9 Active Listening Techniques

TECHNIQUE	DESCRIPTION
Non-verbal	The facilitator uses non-verbal cues such as eye contact and nodding to demonstrate engagement in listening.
Echo	The facilitator echoes or paraphrases the words of the trainees to demonstrate engagement and reinforce their role as a listener.
Reflect	The facilitator provides a reflection on the content of the trainees' discussion.
Short interject	The facilitator very briefly interjects to guide the discussion and maintain direction.

established with respect to the development of high levels of technical skill. From the surgeon perfecting a neat suturing technique, through to the elite golfer perfecting her putting, rehearsal has been the mainstay of enhanced performance, particularly in the domain of psychomotor skills. However, rehearsal unfortunately has not been commonplace with respect to non-technical skills.

Learning will be maximised if immediately after debriefing, the individual or team has the opportunity to try the task again and put in place the strategies discussed in the debrief for enhanced performance. Whether in the same scenario or a variation, the ability to purposefully rehearse is critical.

Unfortunately, due to the constraints of time and resources such as simulator availability, rehearsal has not yet become the dominant mode of learning in non-technical skills training programs. With training, and especially non-technical skills training, occurring quite infrequently in many industries, it is likely that the transfer of training is not optimised, and that the benefits of the training and critical debrief are not nearly as significant as if rehearsal had been built into the training program.

Unpacking the Cognitive 'Frame' of Participants

More recent research has examined specifically the way in which critique of participants' performance can effectively form part of the debriefing process. The early guidance on debriefing emphasised a process of participant-led critical self-reflection. However, this left little room for specific feedback and critique to be provided by the instructor in a more traditional training approach. Especially in environments such as healthcare, where trainees might be less technically proficient and experienced than in the context of recurrent training in aviation, specific feedback and critique from the instructor are warranted.

One debriefing technique that attempts to provide an appropriate framework for instructor-led critique has been termed *debriefing with good judgement*.[35] This advanced debriefing technique, in simple terms, begins with the instructor identifying an area where, in their expert opinion, performance was sub-optimal. The instructor introduces this to the participants and asks participants to reflect on and describe what they were thinking at the time. This is referred to as the instructor eliciting the cognitive frame of the participants. Then, through a process of discussion, the instructor is able to identify any potential deficits in knowledge or understanding of the most appropriate actions and can suggest alternatives that might yield a better outcome in the future.

Debriefing for Learning on the Job

As we have previously emphasised, continuous development of non-technical skills need not only occur through formal training interventions. Rather, the processes of reflective practice, cognitive apprenticeships and other forms of on-the-job skill development are also critical. To this end, debriefing normal day-to-day operations as well as non-normal events and incidents can play an important role in skill development.

Within this context, debriefing is also important to the development of non-technical skills on the job and provides the opportunity for reflection and enhanced learning.

PRINCIPLE TEN: FACILITATED DEBRIEF OF PERFORMANCE IS A CRITICAL ELEMENT OF ASSESSING NON-TECHNICAL SKILLS AND MAINTAINS A CONTINUOUS CYCLE OF TRAINING, ASSESSMENT AND ENHANCED PERFORMANCE

Perhaps the most important aspect of assessing non-technical skills is that assessment supports the facilitated debriefing of performance and as such, makes a significant contribution to learning within a non-technical skills training program. Developing the debriefing skills of instructors and assessors forms an essential component of an effective non-technical skills training program.

References

1. Klampfer, B., Flin, R., Helmreich, R. L., Häusler, R., Sexton, B., Fletcher, G., Field, P. et al., (2001). *Enhancing Performance in High Risk Environments: Recommendations for the Use of Behavioural Markers*. Ladenburg: Daimler-Benz Stiftung.
2. Fletcher, G., Flin, R., McGeorge, P., Glavin, R. J., Maran, N. J., and Patey, R. (2003). Anaesthetist's non-technical skills (ANTS): Evaluation of a behavioural marker system. *British Journal of Anaesthesia*, *90*(5), 580–588.
3. Yule, S., Flin, R., Maran, N., Rowley, D., Youngson, G., and Paterson-Brown, S. (2008). Surgeons' non-technical skills in the operating room: Reliability testing of the NOTSS behavior rating system. *World Journal of Surgery*, *32*(4), 548–556.
4. Seamster, T. L., Redding, R. E., and Kaempf, G. L. (1997). *Applied Cognitive Task Analysis in Aviation*. Aldershot, UK: Ashgate Publishing.
5. Flanagan, J. C. (1954). The critical incident technique. *Psychological Bulletin*, *51*(4), 327.
6. Klein, G. A., Calderwood, R., and MacGregor, D. (1989). Critical decision method for eliciting knowledge. *IEEE Transactions on Systems, Man and Cybernetics*, *19*(3), 462–472.
7. Yule, S., Flin, R., Paterson-Brown, S., Maran, N., and Rowley, D. (2006). Development of a rating system for surgeons' non-technical skills. *Medical Education*, *40*(11), 1098–1104.
8. Linn, R. L., and Gronlund, N. E. (2000). *Measurement and Assessment in Teaching* (8th edn.). Upper Saddle River, NJ: Prentice-Hall.
9. Isaac, S., and Michael, W. B. (1995). *Handbook in Research and Evaluation for Education and the Behavioral Sciences*. San Diego, CA: EdITS.
10. Landy, F. J., and Farr, J. L. (1980). Performance rating. *Psychological Bulletin*, *87*(1), 72.
11. Flin, R., Martin, L., Goeters, K.-M., Hormann, H., Amalberti, R., Valot, C., and Nijhuis, H. (2003). Development of the NOTECHS (Non-Technical Skills) system for assessing pilots' CRM skills. *Human Factors and Aerospace Safety*, *3*, 97–120.
12. Ilgen, J. S., Ma, I. W., Hatala, R., and Cook, D. A. (2015). A systematic review of validity evidence for checklists versus global rating scales in simulation-based assessment. *Medical Education*, *49*(2), 161–173.
13. Phelan, P. (1999). Tailored Training. *Flight Global*(10 February).
14. McDonnell, L. K. (1996). *Facilitation Techniques as Predictors of Crew Participation in LOFT Debriefings*. Ames Research Center Moffett Field, California: National Aeronautics and Space Administration.
15. Balzer, W. K., and Sulsky, L. M. (1992). Halo and performance appraisal research: A critical examination. *Journal of Applied Psychology*, *77*(6), 975.
16. Croskerry, P. (2003). The importance of cognitive errors in diagnosis and strategies to minimize them. *Academic Medicine*, *78*, 1–6.
17. Croskerry, P. (2002). Achieving quality in clinical decision making: Cognitive strategies and detection of bias. *Academic Emergency Medicine*, *9*(11), 1184–1204.

18. Yeates, P., O'Neill, P., Mann, K., and Eva, K. W. (2013). 'You're certainly relatively competent': Assessor bias due to recent experiences. *Medical Education*, *47*(9), 910–922.
19. Eva, K. W., and Regehr, G. (2005). Self-assessment in the health professions: A reformulation and research agenda. *Academic Medicine*, *80*(10), S46–S54.
20. Boet, S., Bould, M. D., Sharma, B., Revees, S., Naik, V. N., Triby, E., and Grantcharov, T. (2013). Within-team debriefing versus instructor-led debriefing for simulation-based education: A randomized controlled trial. *Annals of Surgery*, *258*(1), 53–58.
21. Arora, S., Miskovic, D., Hull, L., Moorthy, K., Aggarwal, R., Johannsson, H., Gautama, S., Kneebone, R., and Sevdalis, N. (2011). Self vs expert assessment of technical and non-technical skills in high fidelity simulation. *The American Journal of Surgery*, *202*(4), 500–506.
22. Pena, G., Altree, M., Field, J., Thomas, M., Hewett, P., Babidge, W., and Maddern, G. (2015). Surgeons' and trainees' perceived self-efficacy in operating theatre non-technical skills. *British Journal of Surgery*, *102*(6), 708–715.
23. Kruger, J., and Dunning, D. (1999). Unskilled and unaware of it: How difficulties in recognizing one's own incompetence lead to inflated self-assessments. *Journal of Personality and Social Psychology*, *77*(6), 1121.
24. Boud, D. (2013). *Enhancing Learning through Self-assessment*. London, UK: Routledge.
25. Savoldelli, G. L., Naik, V. N., Park, J., Joo, H. S., Chow, R., and Hamstra, S. J. (2006). Value of debriefing during simulated crisis management: Oral versus video-assisted oral feedback. *The Journal of the American Society of Anesthesiologists*, *105*(2), 279–285.
26. Dismukes, R. K., McDonnell, L. K., and Jobe, K. K. (2000). Facilitating LOFT debriefings: Instructor techniques and crew participation. *International Journal of Aviation Psychology*, *10*(1), 35–57.
27. McDonnell, L. K., Jobe, K. K., and Dismukes, R. K. (1997). *Facilitating LOS Debriefings: A Training Manual*. Mountain View, CA: Ames Research Center Moffett Field, California, National Aeronautics and Space Administration.
28. Dismukes, K., and Smith, G. M. (2000). *Facilitation and Debriefing in Aviation Training and Operations*. Aldershot, UK: Ashgate Publishing.
29. Ahmed, M., Sevdalis, N., Paige, J., Paragi-Gururaja, R., Nestel, D., and Arora, S. (2012). Identifying best practice guidelines for debriefing in surgery: A tri-continental study. *The American Journal of Surgery*, *203*(4), 523–529.
30. Steinwachs, B. (1992). How to facilitate a debriefing. *Simulation and Gaming*, *23*(2), 186–195.
31. Federal Aviation Administration. (2015). *Advisory Circular 120-35D: Flightcrew Member Line Operational Simulations: Line-Oriented Flight Training, Special Purpose Operational Training, Line Operational Evaluation*. Washington, DC: US Department of Transportation.

32. Dismukes, R. K., Gaba, D. M., and Howard, S. K. (2006). So many roads: Facilitated debriefing in healthcare. *Simulation in Healthcare*, *1*(1), 23–25.
33. Sawyer, T., Sierocka-Castaneda, A., Chan, D., Berg, B., Lustik, M., and Thompson, M. (2012). The effectiveness of video-assisted debriefing versus oral debriefing alone at improving neonatal resuscitation performance: A randomized trial. *Simulation in Healthcare*, 7(4), 213–221.
34. Lyons, R., Lazzara, E. H., Benishek, L. E., Zajac, S., Gregory, M., Sonesh, S. C., and Salas, E. (2015). Enhancing the effectiveness of team debriefings in medical simulation: More best practices. *Joint Commission Journal on Quality & Patient Safety*, *41*(3), 115–125.
35. Rudolph, J. W., Simon, R., Dufresne, R. L., and Raemer, D. B. (2006). There's no such thing as 'nonjudgmental' debriefing: A theory and method for debriefing with good judgment. *Simulation in Healthcare*, *1*(1), 49–55.

6
Instructional Systems Design

Instructional Systems Design

Instructional systems design describes the formal process through which a training program is designed, developed, implemented and evaluated. In many respects, it is the formal process through which the science of instruction, learning and assessment is put into practice to ensure that a training program is deployed effectively and achieves its overall goals.

In any high-risk industry, instructional systems design is critical: a non-technical skills training program can ensure that appropriate skills development takes place and that trainees are able to be assessed as competent in using those skills in practice.

Instructional systems design was first formally described in the 1970s, having emerged from the systematic approach used in various military contexts for the design of training programs in the 1950s and 1960s. At this time, a variety of models were developed to describe requisite activities and stages in the instructional design process. For instance, Dick and Carey described a nine-step model that emphasises the specification of core competencies and the creation of criterion-referenced tests prior to the design of the training program itself.[1] This approach is critical as it defines performance objectives prior to making decisions about the best way to develop the requisite knowledge and skills.

Other models provide either a simpler or a more detailed sequence of steps in the overall process of instructional systems design. However, the majority of these follow a core series of general phases, such as (1) analysis; (2) design and development; (3) implementation; and (4) evaluation (ADDIE).[2]

Over the years, the generic ADDIE model has been criticised for being too narrow in its conceptualisation of how professionals develop, maintain and enhance their skills and for not paying sufficient attention to the less formal ways in which skill development takes place through reflective practice. However, for the purpose of developing non-technical skills programs, we can adapt the generic ADDIE model and include less formal training interventions in our overall design.

Analysis: Identification of Specific Training Needs

The first reason why instructional systems design is important is that it demands that specific training needs are identified and clearly articulated. In many high-risk industries, the need for the training and assessment of non-technical skills is mandated through a regulatory framework. However, often these frameworks demand only a very generic structure for the syllabus and curriculum of non-technical skills training programs. Instructional systems design provides the tools to investigate and identify the specific training needs for a specific cohort of employees with respect to a broad domain such as 'teamwork' or 'situation awareness'.

Design: Specification of Clear Learning Objectives

Instructional systems design demands the specification of the outcomes of a training program. These learning objectives define the knowledge and skills that will be developed through the training program and provide the benchmark against which assessment can be designed. In many industries, these learning objectives take the form of a *competency specification*, which has been described in detail in the previous chapter.

Development: Detailed Curriculum of Training Events and Assessment Tasks

Instructional systems design provides a framework for developing a well-considered syllabus and making appropriate choices with respect to the types of learning activities that can be used to achieve the learning objectives within the competency specification. For each specific training need, an appropriate instructional event and assessment of competency must be created for the development of the core enabling knowledge and constituent skills.

Delivery: Instructional Events in the Form of Training and Assessment

The actual delivery of training and assessment is only one step in the overall ADDIE process of instructional design. The ADDIE process of instructional design emphasises the detailed planning and subsequent evaluation of training that must occur to ensure that the process of developing knowledge, skills and attitudes is systematic and rigorous.

Evaluation of Training Effectiveness

Finally, instructional systems design emphasises that all training programs must be seen within a continuous improvement cycle, and that evaluation of the effectiveness of a training program should inform its on-going development. Evaluation can take a range of forms and use different sources of data in determining the strengths and weaknesses of a non-technical skills training program.

Each of these aspects of instructional system design will be examined in more detail throughout this chapter.

Analysis and Specification of Training Needs

The first critical phase of instructional systems design involves the analysis and specification of training needs. Put simply, this involves developing an understanding of the knowledge, skills and appropriate attitudes required for safe and efficient work performance, identifying the gaps between trainees' current knowledge and skill and that which is needed for safe and efficient performance. Once this analysis has taken place, the next step is to clearly specify what remains to be developed through the formal training program.

Risk-Based Training Needs Analysis

The term *training needs analysis* refers to the basic process of identifying gaps between current knowledge, skills and appropriate attitudes and those required for satisfactory job performance. Recent work within the rail industry, particularly in the UK, has extended the scope of training needs analysis and has incorporated the concept of risk into the decision-making around training design.[3] Called *risk-based training needs analysis* (RBTNA), this approach shares an initial commonality with traditional training needs analysis through the undertaking of a task analysis as the first stage. However, RBTNA differs from traditional training needs analysis through the next stage, in which risks associated with each tasks are assessed.

The approach of RBTNA has three main phases. First, a process of *role definition* is undertaken, which involves a detailed task analysis, and for each task element, the identification of requisite knowledge and skills. This includes both technical and non-technical knowledge and skills. Second, a process of *assessing training priority* is undertaken and is based around the assessment of risk. At this stage, the hazards associated with the job are analysed, and a detailed assessment of the likelihood and consequence associated with sub-optimal task performance is quantified, along with an analysis of the frequency and difficulty of the specific task. Finally, an analysis of training options is undertaken to inform final training design.[4]

Although the RBTNA methodology has been to a large degree constrained to the rail environment, it has significant potential as a training needs analysis methodology across all high-risk environments and has significant strength in the way in which it integrates the analysis of both technical and non-technical knowledge and skill.

Cognitive Task Analysis

Another approach to the identification of training needs for non-technical skills is cognitive task analysis (CTA). As described in the previous chapter, CTA was developed from traditional task analysis and is a method used to describe the mental processes that support work performance alongside the behavioural elements.[5] The processes for undertaking cognitive task analysis include methods such as cognitive interview techniques, verbal report methods in which the expert operators describe what they are doing/thinking, and process mapping information, communication and decision pathways.

Cognitive Work Analysis

Traditional approaches to training needs analysis have been criticised for the way in which they focus on the individual operator's tasks and in doing so, fail to capture the broader demands of managing off-nominal events in the highly dynamic and complex environments of high-risk industries.[6] Traditional training needs analysis, which focusses on developing a taxonomy of operator actions for specific work practices, is ideally suited to routine operations under normal conditions. Here, an operator's work can be specified in a detailed procedure, and the knowledge and skill requirements to perform that procedure can be relatively easily laid out.

However, with the advent of increasingly complex systems of work, where different levels of automation are deployed, the operator can shift rapidly from a system monitoring mode into a much more dynamic problem-solving mode when abnormal conditions arise. For instance, the work of an anaesthetist in inducing and maintaining anaesthesia can be readily subjected to task analysis, a step-by-step procedure can be developed to describe those tasks, and the requisite knowledge and skills can be specified. However, the analysis of work performed during an anaesthetic crisis, with equipment malfunction occurring concurrently with rapid deterioration in the patient's condition, is much less amenable to traditional task analysis. To this end, cognitive work analysis has been put forward as an alternative methodology.

For many, the process of cognitive work analysis is largely impenetrable due to its complexity, and as such, it is not a technique that is readily deployed by anyone other than experts in the methodology. It is far beyond the scope of this book to describe in detail the processes of cognitive work analysis. These have been laid out in detail in several foundational texts.[7,8] However, the general principle of thinking beyond task analysis for normal system operation is a useful concept to bear in mind when developing non-technical skills programs.

Addressing Organisational and Individual Needs

Any non-technical skills training program is significantly more powerful in terms of enhancing performance and safety if it is *responsive* to both the strengths and the weaknesses of an organisation, as well as individual priorities for skill development.

The modern non-technical skills training program is now best seen as an *integrated* part of an organisation's on-going efforts to continually improve performance. In this formulation, there are a number of specific interfaces between the non-technical skills training program and other organisational systems such as safety audits and investigations, safety reports, and the regular review of international trends in safety occurrences and training interventions.

CASE STUDY: LOSS OF CONTROL IN-FLIGHT (LOCI)

In recent years, there has been a spate of catastrophic aviation accidents caused by flight crew losing control of the aircraft in flight. Infamous accidents such as Air France 477, Colgan Air 3407 and Air Asia 8501 are all examples of this category of accident, where flight crew have been surprised by a non-normal situation and have failed to accurately diagnose and respond to the situation to effectively recover. Over the last decade, this form of accident has become a critical issue facing the industry worldwide and has surpassed other common accident categories such as controlled flight into terrain (CFIT) with respect to the number of accidents and fatalities.[9]

An analysis of these types of accident revealed that flight crew frequently respond to non-normal events with a 'startle effect', which in turn has significant impacts on their effective use of non-technical skills such as situation awareness, problem diagnosis and communication. These performance deficiencies in turn prevent effective recovery actions being implemented by flight crew.[10]

In response to these findings, research has been undertaken to evaluate the effectiveness of traditional approaches to training for the types of events that precipitate LOCI accidents, such as high-altitude stall. The traditional approach to training such events involves a simulator session in which the flight crew revise for non-normal situations and are briefed by an instructor to refresh the correct aircraft handling techniques and standard operating procedures prior to encountering the situation in the simulator.

This training has been criticised for relying on highly scripted and predictable exercises. Moreover, these exercises focus more on technical skills rather than demanding that the flight crew draw on their suite of non-technical skills. Recent research has highlighted that this traditional approach is associated with significantly better flight crew performance, due to the effects of expectation, when compared with situations where the flight crew are presented with the non-normal situation under less predictable circumstances, as might occur during a real flight.[11]

This case study of LOCI accidents provides an exemplar of how the traditional forms of training in many high-risk industries fail to

- Adequately learn from recent accidents to identify emerging industry needs
- Have training systems that are sufficiently flexible to incorporate such learning
- Adopt instructional strategies that integrate the technical and non-technical aspects of performance

Recent advances towards data-driven and evidence-based training approaches provide significant potential benefits to enhancing performance and safety through effective non-technical skills training programs.

Evidence-Based Training

The term *evidence-based training* is frequently used to describe training programs that are designed to respond directly to the needs of the industry, the organisation and the individual. An early example of this approach to training curriculum design can be traced to the development of the Advanced Qualification Program (AQP) in commercial aviation.[12]

In the decades prior to the advent of AQP, already certified flight crew were required to undergo recurrent training (usually annually) that stepped through a formal syllabus set out in national regulations. This approach was designed to ensure that a defined set of critical flight manoeuvres were regularly rehearsed in a highly structured and prescriptive fashion, such that flight crew competence could be demonstrated. While this approach gave some reassurance as to a flight crew member's ability to perform this set of critical flight manoeuvres, it provided for a 'one size fits all' approach to flight crew training that did not leave room for any tailoring to specific organisational or individual needs.

As a voluntary alternative to the prescriptive recurrent training curriculum, AQP was designed to provide proficiency-based training and evaluation that formally included the training and assessment of non-technical skills. Moreover, AQP sought to draw together the methods used for the design and implementation of training for all safety-critical personnel such as pilots, flight engineers, flight attendants, aircraft dispatchers, instructors, evaluators and other operations personnel. Of note is the fact that AQP focussed on the crew, rather than the individual, as the focus of training, and emphasised team processes in both normal and non-normal situations.

Undoubtedly, the most important feature of AQP was that it was explicitly *data-driven* in nature, such that an airline had to draw on a variety of sources of data and follow a formal process of instructional systems design in the development of its training programs.

While AQP did not perhaps have the level of uptake desired of such an innovative approach to training design, recent efforts in commercial aviation have reignited the philosophy of AQP in the more modern guise of *evidence-based training*.

In 2007, the International Air Transport Association (IATA) launched the IATA Training and Qualification Initiative (ITQI), a keystone program of which was evidence-based training (EBT).[13] The EBT approach was designed to apply the principles of competency-based training and develop a global competency framework for different generations of commercial aircraft based on needs identified from data sources including flight data analysis, observations of normal flight operations, accident and incident investigations, and other sources of air safety data. The EBT program is an excellent example of an industry-wide approach to training needs analysis and competency specification.

Data from the Safety Management System

Across all high-risk industries, the systematic management of safety is a key feature of organisational efforts to reduce the likelihood of catastrophic events. Accordingly, the majority of organisations in high-risk industries have a safety management system (SMS) that sets out a series of organisational processes for the identification of hazards, mitigation of risk, measurement of performance, investigation of incidents and on-going continuous improvement.

One of the easiest ways of ensuring that a non-technical skills training program addresses an organisation's needs is to use the SMS as a source of data for training needs analysis and also a source of data for evaluating the effectiveness of non-technical skills training programs.

In the last decade or so, different sources of data have been developed as parts of an organisation's SMS. Of these, the process of 'normal operations monitoring' has provided some significant benefits, enabling an organisation to gain an understanding of what really happens during normal everyday work and how well the organisation is supporting its operators through procedures, equipment, and training. Originally developed by the University of Texas as Line Operations Safety Audits (LOSA) for use in commercial aviation,[14,15] the methodology has been adapted for use in a range of other industries.[16] The evaluation of performance across the domains of non-technical skills has from the very outset been a focus of normal operations monitoring. To this end, it has become a very important component of training needs analysis for non-technical skills programs in high-risk industries. Table 6.1 provides an overview of the 10 defining characteristics of normal operations monitoring.

Gap Analysis

The final stage in the process of training needs analysis is to undertake a gap analysis, which is designed to understand the needs of the trainees in terms of their current knowledge, skills and attitudes, and what is required for competent performance in their safety-critical roles.

This has been an area where non-technical skills training programs have often failed to be thorough, and as a consequence, have been

Table 6.1 Characteristics of Normal Operations Monitoring

CHARACTERISTIC	DESCRIPTION
Observations are made during normal operations	The focus of the method is to gain an understanding of what actually occurs and to understand the strengths and weaknesses of day-to-day operations.
Teams volunteer to participate	There is no obligation for anybody to be observed in their day-to-day work. One of the important characteristics is that people feel comfortable being observed.
Data collection is anonymous, confidential and safety-minded	To gain a sense of normal operations, anybody who is observed must entirely trust in the fact that their performance will remain strictly anonymous and confidential. In this way, people will feel comfortable to perform tasks as they normally would, instead of 'doing it by the book' because they are being watched.
Joint sponsorship between management and the workforce	As part of developing this trust around the data being strictly anonymous and confidential, the data collection needs to be endorsed by management and workforce representatives.
The project has appropriate targets for data collection	The project needs to be designed such that sufficient data are collected to be able to make meaningful conclusions.
The program employs trusted and trained observers	The people undertaking observations and collecting the data need to be trusted by the workforce and also trained in the observation and data collection methodology.
The program has a trusted data collection site	The workforce needs to feel comfortable that the raw data will not be unduly scrutinised by supervisory staff and management. Often, an independent organisation is used to facilitate data collection and analysis.
Data are scrutinised before analysis	A working group needs to scrutinise all data prior to analysis to ensure that valid conclusions can be drawn and any biases or errors in data collection can be removed.
The results are given to the workforce	To maintain trust and transparency, the results of the project need to be provided to the whole workforce.
Data are used to identify targets for enhancement	The sole aim of normal operations monitoring is to identify the strengths and weaknesses of normal operations and to identify targets for enhanced safety.

Source: ICAO, *Line Operations Safety Audit (LOSA)*, Montreal, Canada, International Civil Aviation Organisation, 2002.

criticised for being overly generic and failing to provide adequate return on investment. The main consequence of not performing a gap analysis is that trainees do not value the training, as it does not address an area of knowledge or skill deficit relevant to their work performance.

If, prior to designing a non-technical skills training program, an organisation has not been able to demonstrate a critical training need, the training program is unlikely to contribute to enhanced safety and productivity in the workplace.

Personalised Training

Beyond the use of industry-wide and organisational data sources to develop an evidence-based non-technical skills training program, there is considerable merit in also developing training that responds to individual needs for skill development and enhancement. This is an important characteristic of on-going professional development and should not be ignored in the design of non-technical skills training programs.

Another aspect of personalising training is the need to consider individual trainees' preferred learning styles, as described in Chapter 3. Administering a learning styles inventory, such as Kolb's,[18] at the commencement of a training program can give instructors insights into how they can best tailor learning experiences to maximise knowledge and skill development.

Defining Curriculum Outcomes and Instructional Objectives

The next stage after the task analysis is complete is to carefully set out the outcomes of the training program and the detailed learning objectives. Since as early as the 1940s, the importance of setting clearly defined training objectives has been known, and such objectives become the criteria for all subsequent aspects of instructional design, including mapping out the syllabus of content, choosing instructional techniques, and designing appropriate assessment tools.[19] The training objectives set out the outcomes against which all aspects of instructional design must serve. The literature on instructional systems design is unfortunately somewhat confusing with respect to the terminology used here, with the terms *goals*, *aims*, *objectives* and *outcomes* used somewhat interchangeably. For the purposes of this book, we will adopt a simple convention with respect to these terms.

First, we will refer to the overarching products of the non-technical skills training program as *curriculum outcomes*, and these will encompass the broad aims and objectives with respect to operationally enhancing safe and efficient performance. Second, we will use the term *instructional objectives* to describe the more detailed aspects of knowledge, skill and attitudinal development at a higher level of specificity. This is an approach adopted in medical education, where a predominant focus on very highly detailed instructional objectives was leading to the loss of a bigger picture of the overall curriculum.[20] Instructional objectives, as descriptors of the products of training at a micro-level, do not easily capture the integrated whole that is professional practice in a high-risk industry. Relying on instructional objectives alone can in turn lead to a fragmented view of training design that lacks integration and overall purpose. It has been suggested that curriculum outcomes should be written in a manner that is intuitive and user friendly. They should be able to be used easily in curriculum planning, in teaching and in assessment.[20]

While detailed instructional objectives are used primarily by training staff in the process of instructional systems design, the broader curriculum objectives have a much wider audience. First, within the typical non-technical skills training program, a group of facilitators will be responsible for the delivery of the program. For this group, a coherent set of user-friendly curriculum objectives serves an important role in defining the outcomes of the training, rather than the highly detailed instructional objectives. As discussed shortly, the audience for instructional objectives is the instructional design team within the training department, who use them in making instructional decisions, creating lesson plans, and designing case studies, simulator scenarios and assessment tools.

Further, the curriculum objectives are important in 'selling' the investment in training to management. Across most high-risk industries, training is significantly constrained by budget, and a set of user-friendly curriculum objectives should lay out the essence of the business case for the training program. Table 6.2 outlines the different aspects of curriculum outcomes and instructional objectives in terms of scope, audience and purpose.

Specifying Curriculum Outcomes

Curriculum outcomes can be specified at the programmatic level, describing the overarching aims and objectives of an organisation's

Table 6.2 Aspects of Curriculum Outcomes and Instructional Objectives

	SCOPE	CORE AUDIENCE	PURPOSE
Curriculum outcomes	Broad user-friendly outline of integrated outcome of training	Management Instructors Facilitators Trainees	Business case Safety impact
Instructional objectives	Detailed specification of knowledge, skills and attitudes to be developed from training	Instructional designers Training department	Curriculum design Lesson planning Case study design Scenario development Assessment design

non-technical skills training program. Alternatively, they can be specified with respect to a small element of an organisation's non-technical skills training program, such as the recurrent training program for flight crew, or the induction program for maintenance engineers.

The development of curriculum outcomes should in many ways simply involve stating the overall goals of the training program. They are by their very nature less prescriptive in formulation than instructional objectives and need to capture the essence of the training program in response to an identified need.

For a complete introductory non-technical skills training program, the curriculum outcomes might begin with a 'top-level' statement describing the overall aim of the training. This could then be broken down into more specific statements of curriculum outcome in each of the domains of non-technical skills. Table 6.3 provides some examples of top-level curriculum objectives.

An additional benefit of being explicit in describing curriculum outcomes is that they provide the top-level taxonomic categories for the non-technical skills training program. These can then be elaborated on in more detail in the form of very specific instructional objectives.

Specifying Instructional Objectives

Instructional objectives are the highly detailed specifications of what a trainee *should be able to do* at the completion of training. Originally developed within a behaviouristic frame for the specification of technical skills, instructional objectives are typically anchored around

Table 6.3 Example 'Top-Level' Curriculum Outcome Statements from Non-Technical Skills Training Programs

	TOP-LEVEL CURRICULUM OUTCOME STATEMENTS
Early aviation crew resource management[21]	To train aircrews to use all available resources – equipment, people, and information – by communicating and coordinating as a team.
Anaesthesia crisis resource management[22]	To address many non-technical individual and team performance skills important for improved patient safety, as well as issues related to resident supervision.
	To understand and improve participants' proficiency in crisis resource management (CRM) skills and to learn skills for debriefing residents after critical events.
Oil and gas well operations teams[23]	To improve the skills of the individual worker in a team setting and address behaviour in routine operations with the aim of avoiding critical incidents.
Rail resource management[24]	The objective of rail resource management is to ensure that front-line operators possess the competencies necessary to perform safely in all circumstances.

a specific observable behaviour and its performance outcomes. For instance, in the *ab initio* flight training context, an instructional objective might be anchored around the observable behaviours associated with maintaining straight and level flight, such as trimming the aircraft and responding to altitude deviations in a prompt and smooth fashion.

Instructional objectives typically have three components: (1) a performance statement; (2) the conditions in which that performance will occur; and (3) the standards to which the performance will be judged as acceptable.[25] With respect to the example provided in the previous paragraph, a basic instructional objective would be:

> The pilot shall maintain straight and level flight (performance), in Instrument Meteorological Conditions (IMC) and moderate turbulence (conditions), within plus or minus 50 feet of the cleared altitude (standards).

This instructional objective is quite easy to specify, as the performance is easily observable, as are the standards against which it is measured. It could be suggested that such clearly specified instructional objectives would be much more difficult to create for performance within non-technical domains such as communication or situation awareness. However, with some careful consideration and reference to task analyses, as discussed before, it is not that difficult to create instructional objectives for non-technical skills training programs. We will explore in more detail the notion of moving towards less generic frameworks for non-technical skills in future chapters. However, at this point, it is worth noting that the process of writing instructional objectives urges us to think about non-technical skills in a situated manner, in the contexts in which they will be deployed.

Performance Statements

Specifying performance statements for non-technical skills adopts a similar approach to that used in the more traditional technical domains. Performance statements by no means need to be restricted to behaviours within the domain of observable actions on elements of the environment. Rather, performances with respect to knowledge, skills and attitudes can all be described. Similarly, returning to our domains of learning described in Chapter 3, it is also possible to describe performance statements within the cognitive, affective and psychomotor domains. A very useful tool for developing performance statements across the various different domains of learning is to develop a set of *verbs* that can be used as the primary descriptor of the performance. Similarly, performance statements do not need to be restricted to *overt* (directly observable) behaviours. Rather, *covert* behaviours, such as those found within the cognitive domains of non-technical skills such as situation awareness and decision-making, can also be the targets of performance statements. Table 6.4 provides a few exemplar verbs that can be used in non-technical domains.

Conditions The conditions within which non-technical performance is required to occur may be many and varied in any high-risk industry environment. The most important aspect is to consider performance during both normal and non-normal operations. One of the characteristics of high-risk industries is that safety ultimately rests in the ability of the operators to recover from emergency situations.

Table 6.4 Sample Verbs for Non-Technical Skills Performance Statements

SITUATION AWARENESS	DECISION-MAKING
Observe	Identify
Notice	Evaluate
Comprehend	Consider
Anticipate	Weigh up
COMMUNICATION	TASK MANAGEMENT
State	Plan
Inquire	Prioritise
Suggest	Delegate
Prompt	

Therefore, conditions for non-technical performance will vary from everyday 'business as usual' through to crisis situations.

The conditions will invariably be task dependent, but consideration also needs to be given to the social and cultural context in which everyday work activity takes place. Specifying the conditions under which skills will be developed and under which performance will be assessed lends a great deal to the next stage of curriculum development, and in particular to the process of scenario design, if simulation is to be used in the non-technical skills training program.

Standards The point at which instructional objectives for non-technical skills become more difficult is where standards for acceptable performance need to be specified. Unlike the example of maintaining straight and level flight, where a flight instrument provides the instructor with a constant readout of the standard of performance (the altimeter), similar continuous quantitative measures are very rarely provided with respect to performance in non-technical skills domains.

However, the fact that the non-technical skills training programs we are exploring in this book are explicitly focussed on achieving safe and efficient operations in high-risk industries provides us with some clues as to how we might tackle the relative lack of quantitative measures of performance. High-risk industries are defined in terms of *safety* and *risk*, and aspects of performance such as *timeliness* can be used to set performance standards. While these might align more with the *outcomes* of non-technical performance, they still provide appropriate measures.

Curriculum Design

The process of actually putting together a non-technical skills training program is referred to as *curriculum design*. This stage of instructional systems design specifies the types of learning activities that will take place and also what resources will be required to deliver the training. This process can be undertaken effectively only once training needs have been assessed and the broad curriculum outcomes and specific learning objectives have been defined.

The broad process of curriculum design involves working backwards from the competency specification and learning objectives. In sequence, assessment tasks are designed, and then decisions are made with respect to the designing of training events, appropriate sequencing of these events, specifying the materials and equipment needed to support the training event, and the creation of guidance material for instructors and trainees. Figure 6.1 provides an outline of this process.

The following sections will briefly explore the pertinent considerations relating to each of these steps as they relate to a non-technical skills training program.

Designing Assessment Tasks

Almost counter-intuitively, the first elements of the curriculum that should be designed are the forms of assessment that will be used to measure trainees' competency development. Designing assessment tasks first serves to focus the overall curriculum design on the desired knowledge and skills that are to be developed. Working backwards, it is actually a natural progression from the process of competency specification and definition of learning objectives to the design of

1. *Designing assessment*
2. *Designing training events*
3. *Sequencing training events*
4. *Specifying equipment and materials*
5. *Generating guidance material for instructors and trainees*

Figure 6.1 Stages in curriculum design.

assessment. The appropriate assessment of non-technical skills, including the core enabling knowledge, constituent skills and overall performance, has been explored in detail in the previous chapters.

Designing Training Events

Designing for the development of core enabling knowledge: Once decisions have been made with respect to assessment, the next stage of curriculum design involves planning the training events through which trainees will develop the required core enabling knowledge. We have already examined in detail a range of training modes that could be used for knowledge development, including the traditional seminar presentations, case study discussions and self-directed eLearning packages.

Designing for skill development: Once core enabling knowledge has been developed, the next task is to plan the training events that will be used for developing and practising actual non-technical skills. In Chapter 5, we explored in detail the modes of training that can be used for non-technical skill development, including the role-play, simulation and on-the-job forms of training.

Sequencing Training Events

No training event can be seen in isolation, and as with any training program, consideration needs to be given to progressive development of knowledge and skills and the on-going opportunity to practise these skills for continued competence and reinforcement.

Specifying Equipment and Materials

Another important consideration in curriculum design involves specifying and sourcing the equipment and materials required to deliver the training. Due to their nature, activities such as role-play and especially simulation require a significant investment in resources to ensure that an adequate degree of fidelity is achieved and that trainees are exposed to an authentic learning experience.

Generating Guidance Material for Instructors and Trainees

A final aspect of curriculum design involves generating guidance materials for both instructors and trainees. These might take the forms of notes, scripts, and assessment and feedback forms. The organisation and presentation of these resources are an important consideration for delivering high-quality and professional non-technical skills training.

Implementation and Delivery

The previous chapters have explored in detail a range of considerations relating to the implementation and delivery of non-technical skills training programs and provided practical guidance for effective training and assessment. However, as we are considering the overarching process of instructional design, a few remaining points can be made.

Learner Engagement

The first point to consider when implementing and delivering a training program is learner engagement. As a characteristic of adult learning, trainees need to be forewarned with respect to the scope and benefits of undertaking a training event that focusses on non-technical skills. Much of this can be provided prior to the delivery of training as a pack of briefing materials. The more prepared the learner is for the training, the better the chance of maximising learning outcomes.

Instructor Preparation

The final determinant of whether a non-technical skills training program will be a success rests with the degree of preparedness of the instructional staff. As discussed in Chapter 4, considerable investment in the training of instructors and assessors is required for non-technical skills training. It is not sufficient for an instructor to simply be a subject matter expert. Rather, they require a range of additional skills relating to instructional technique and assessment. Specifically with respect to non-technical skills training programs, skills in facilitated debrief and in the use of behavioural marker systems are absolutely critical.

Evaluation of Training Programs

The final aspect of instructional systems design involves the evaluation of training programs. As discussed previously, the overall framework of instructional systems design suggests that continuous improvement cycles should inform the way in which a training program evolves.

The evaluation of a non-technical skills training program serves a number of important purposes. First and foremost, evaluation should enable the training designers to demonstrate that the objectives of the training program are being met, and that the training needs are being appropriately addressed. This is obviously critical in high-risk industries, where it is essential that the skills required by individuals and teams for safe and efficient performance are appropriately developed. Additionally, evaluation can provide information about how trainees enjoyed the training and what areas of the training program could be enhanced.

One of the most popular frameworks for evaluation was developed by Donald Kirkpatrick in the 1970s and has been used extensively in the evaluation of training programs since.[26] This framework sets out four major levels of evaluation: (1) participant reactions to training; (2) assessment of learning achieved from training; (3) behavioural change after training; and (4) the organisational results achieved from the training.

The first level, that of participants' *reactions*, sets out to ascertain whether trainees enjoyed the training, whether they found the training useful and applicable to their work, and whether they thought they had developed the anticipated knowledge and skills from the training program. This information is useful in determining whether the training program was engaging and whether it met trainees' perceived needs. Published studies have consistently found that trainees have positive reactions to non-technical skills programs, especially when they include more active modes of learning.[21]

The second level, that of *learning*, sets out to measure the degree to which trainees developed the knowledge, skills and attitudes as per the objectives of the non-technical skills training program. This information is critical in evaluating whether the training program as delivered is meeting the learning objectives specified during the training needs analysis. Initially, the measurement of attitudinal change was the most common form of evaluation at this level.[21] However, more and more industries are now using formal assessment of knowledge developed in relation to applied human factors issues as well as assessment of non-technical performance through the use of behavioural markers, as discussed in the previous chapter.

The third level of evaluation, that of *behavioural change*, sets out to measure the degree of transfer of training from the training program to everyday operations. This level of evaluation is critical in determining whether the non-technical skills training program has the desired effects with respect to the safety and efficiency of operations. In the majority of high-risk industries, operators are required to undertake periodic requalification and are subjected to regular checks of competency. While these checks have traditionally been technical in nature, the development of behavioural marker systems now enables such assessment to include non-technical skills.

The fourth and final level of evaluation, that of *organisational impacts*, sets out to measure whether the enhanced knowledge and skills developed in the program are having an effect on the performance of the organisation. This level is critical in assessing the return on investment of non-technical skills training programs. The organisational impact of non-technical skills training programs on organisational performance relates to the overall reduction in incidents and accidents, and enhanced safety and efficiency of operations. Given the infrequent occurrence of poor safety outcomes in high-risk industries, collecting data at this level is extremely difficult.[21] However, the aspects of an organisation's safety management system that can be used in the identification of training needs, as described previously in this chapter, can also be used to monitor the effectiveness of non-technical skills training programs. The on-going analysis of trends in organisational data, such as normal operations monitoring, analysis of incident reports and incident investigations, can identify areas where improved non-technical performance has occurred. Conversely, these forms of evaluation data are also able to identify areas where there have been instances of sub-optimal non-technical performance and identify areas for further training.[27]

Table 6.5 The Four Levels of Evaluation and Sources of Data

LEVEL OF EVALUATION	SOURCES OF DATA
REACTION What a trainee likes and dislikes about the non-technical skills training program.	TRAINEE FEEDBACK • Surveys/questionnaires
LEARNING The degree to which trainees developed the knowledge, skills and attitudes as set out in the objectives of the non-technical skills training program.	TRAINEE ASSESSMENT • Formative assessment of knowledge and skill • Summative tests of knowledge • Assessment of performance using behavioural markers • Safety attitudes surveys
BEHAVIOURAL CHANGE The degree to which the learning in the training program is transferred to improved non-technical performance in everyday work performance.	WORKPLACE OBSERVATIONS • On-the-job assessments • Checks/audits • Normal operations monitoring (e.g. LOSA)
ORGANISATIONAL OUTCOMES The ways in which the training program influences organisational performance with respect to enhanced safety.	SAFETY MANAGEMENT SYSTEM • Reduction in incidents and accident metrics • Safety culture surveys

Source: Adapted from Kirkpatrick, D.L., *Training and Development Handbook: A Guide for Human Resource Development*, New York, McGraw-Hill, 1976.

Overall, the evaluation process for non-technical skills training programs will involve a cycle of continuous improvement, in which the evaluation data have identified emerging training needs, and these needs have been responded to through a training system that enhances overall operational performance.[28] A summary of the four levels of evaluation in Kirkpatrick's model is provided in Table 6.5.

References

1. Dick, W., and Carey, L. (1978). *The Systematic Design of Instruction*. Glenview, IL: Scott, Foresman and Company.
2. Visscher-Voerman, I., and Gustafson, K. L. (2004). Paradigms in the theory and practice of education and training design. *Educational Technology Research and Development*, *52*(2), 69–89.
3. Pitsopoulos, J., and Luckins, R. (2012). Leading practice in rail training and competence management. In J. R. Wilson, A. Mills, T. Clarke, J. Rajan and N. Dadashi (Eds.), *Rail Human Factors around the World* (pp. 728–736). Boca Raton, FL: CRC Press.

4. Shah, P., Taylor, A., and Bonsall-Clarke, K. (2013). The development of a risk-based training needs analysis methodology and tool. In N. Dadashi, A. Scott, J. R. Wilson and A. Mills (Eds.), *Rail Human Factors: Supporting Reliability, Safety and Cost Reduction*. London, UK: CRC Press.
5. Seamster, T. L., Redding, R. E., and Kaempf, G. L. (1997). *Applied Cognitive Task Analysis in Aviation*. Aldershot, UK: Ashgate Publishing.
6. Naikar, N., and Sanderson, P. (1999). Work domain analysis for training-system definition and acquisition. *The International Journal of Aviation Psychology*, *9*(3), 271–290.
7. Rasmussen, J., Pejtersen, A. M., and Goodstein, L. P. (1994). *Cognitive Systems Engineering*. New York, NY: Wiley.
8. Vincente, K. (1999). *Cognitive Work Analysis*. Mahwah, NJ: Lawrence Erlbaum Associates.
9. International Air Transport Association. (2015). Loss of Control In-Flight Accident Analysis Report First *Edition 2010–2014*. Montreal, Canada: IATA.
10. Martin, W. L., Murray, P. S., Bates, P. R., and Lee, P. S. (2015). Fear-potentiated startle: A review from an aviation perspective. *The International Journal of Aviation Psychology*, *25*(2), 97–107.
11. Casner, S. M., Geven, R. W., and Williams, K. T. (2012). The effectiveness of airline pilot training for abnormal events. *Human Factors*, *55*(3), 477–485.
12. Federal Aviation Administration. (1991). *Advisory Circular 120-54: Advanced Qualification Program*. Washington, DC: US Department of Transportation.
13. International Civil Aviation Organization. (2013). *Manual of Evidence-based Training: Doc 9995 AN/497*. Montreal, Canada: ICAO.
14. Helmreich, R. L. (2002). Managing threat and error in aviation and medicine. In *Proceedings of the Third LOSA week* (pp. 15–22). Dubai, United Arab Emirates: International Civil Aviation Organization.
15. Klinect, J. R. (2002). LOSA searches for operational weaknesses while highlighting systemic strengths. *International Civil Aviation Organization (ICAO) Journal*, *57*(4), 8–9, 25.
16. McDonald, A., Garrigan, B., and Kanse, L. (2007). Confidential observations of rail safety (CORS): An adaptation of line operations safety audit (LOSA). In J. M. Anca (Ed.), *Multimodal Safety Management and Human Factors*. Aldershot, UK: Ashgate Publishing.
17. ICAO. (2002). *Line Operations Safety Audit (LOSA)*. Montreal, Canada: International Civil Aviation Organisation.
18. Kolb, D. (1976). *Learning Styles Inventory*. Boston, MA: McBer.
19. Tyler, R. W. (1949). *Basic Principles of Curriculum and Instruction*. Chicago, IL: University of Chicago Press.
20. Harden, R. M. (2002). Learning outcomes and instructional objectives: Is there a difference? *Medical Teacher*, *24*(2), 151–155.
21. Salas, E., Burke, C. S., Bowers, C. A., and Wilson, K. A. (2001). Team training in the skies: Does crew resource management (CRM) training work? *Human Factors*, *43*(4), 641–674.

22. Blum, R. H., Raemer, D. B., Carroll, J. S., Sunder, N., Felstein, D. M., and Cooper, J. B. (2004). Crisis resource management training for an anaesthesia faculty: A new approach to continuing education. *Medical Education*, *38*(1), 45–55.
23. Flin, R., Wilkinson, J., and Agnew, C. (2014). *Crew Resource Management for Well Operations Teams*. London, UK: International Association of Oil & Gas Producers.
24. Lowe, A. R., Hayward, B. J., and Dalton, A. L. (2007). *Guidelines for Rail Resource Management*. Fortitude Valley, Australia: Rail Safety Regulators' Panel.
25. MacLeod, N. (2001). *Training Design in Aviation*. Aldershot, UK: Ashgate Publishing.
26. Kirkpatrick, D. L. (1976). Evaluation of training. In R. L. Craig (Ed.), *Training and Development Handbook: A Guide for Human Resource Development* (2nd edn., pp. 18.11–18.27). New York: McGraw-Hill.
27. Thomas, M. J. W. (2004). Predictors of threat and error management: Identification of core non-technical skills and implications for training systems design. *International Journal of Aviation Psychology*, *14*(2), 207–231.
28. O'Connor, P., Campbell, J., Newon, J., Melton, J., Salas, E., and Wilson, K. A. (2008). Crew resource management training effectiveness: A meta-analysis and some critical needs. *International Journal of Aviation Psychology*, *18*(4), 353–368.

7

Training and Assessing Situation Awareness

Situation Awareness: A Primer

Situation awareness is one of the most important domains of *cognitive* non-technical skills. It describes the processes involved in maintaining an understanding of the status of the whole work environment. Degraded situation awareness has been implicated in many catastrophic events throughout the history of high-risk industries.

Situation awareness has been formally defined as 'the perception of elements in the environment within a volume of time and space, the comprehension of their meaning, and the projection of their status into the future'.[1] This model by the pioneer of research in the domain, Dr Mica Endsley, sets out three main components of situation awareness: (1) *perception* of the current situation; (2) *comprehension* of the current situation in terms of system state; and (3) *projection* of the current situation into the future.[2]

This construction of situation awareness presents a broad functional model of our conscious engagement with the work environment and describes the way in which we develop a mental picture of a situation prior to taking action. Situation awareness in this way can be seen as the emergent feature of the skills associated with situation assessment in combination with expertise and the maintenance of a detailed mental model of both the desired and the actual state of a system.

Situation awareness is also a unifying theory inasmuch as it brings together quite a large array of cognitive processes to describe our overarching consciousness. To some degree, it is a 'catch-all' functional model rather than a detailed empirical description of each of the detailed cognitive processes. It is this simple explanatory power that perhaps has made the concept of situation awareness that much more appealing at the practical level than more complex models of human information processing.

The model highlights several important concepts of situation awareness. First, the three levels of situation awareness form part of a cyclic process of situation awareness, decision and action in a continuous loop of sensing, understanding, acting and re-evaluating. Second, a range of *individual factors* influence situation awareness and the decision-making process, and these factors represent important general aspects of human information processing. Third, a range of *task and system factors* influence situation awareness and the decision-making process.[2]

Situation awareness is frequently defined in the negative through the use of the terms *degraded*, *poor* or *loss of situation awareness*. Each of these terms represents instances where an operator has sub-optimal situation awareness. Conversely, optimal situation awareness is not a single or definable state; rather, it is an extremely task- and context-specific construct. This is because in some situations, the cues and sources of information important for a complete mental model of the situation may be quite different from those needed in another situation. Therefore, we will refer to the term *optimal situation awareness* as our ideal aim point.

On a cautionary note, there is a school of thought that suggests the concept of situation awareness is more of a folk model than a domain of non-technical skill.[3] Associated with this criticism is a further criticism that the domain might be too broad to be sufficiently meaningful, and that it may not be in any way a measurable construct.[4] While it is true to say that our understanding of the complex gestalt of conscious awareness cannot be readily explained by such a broad concept as situation awareness, this chapter holds that there still remains a meaningful and concrete construct that can be the focus of training and assessment.

Situation Awareness: Exemplar Case Study

One of the most recent examples of an accident significantly associated with situation awareness is the loss of Air France Flight AF447. The accident sequence began with incorrect speed indications and the disconnection of the autopilot and auto-thrust due to critical sensors on the outside of the aircraft becoming blocked with ice crystals.

In the following minutes, the crew were unable to accurately comprehend the situation and take the required actions to maintain the aircraft within the parameters of safe flight. The flight crew lost control of the aircraft, which subsequently suffered an aerodynamic stall and crashed into the ocean.[5]

A number of issues related to situation awareness were present in the accident. First, the flight crew did not respond to auditory warnings about the aerodynamic stall. This suggests that the crew may not have *perceived* the warnings, which is a possibility due to auditory insensitivity and the dominance of visual perception in this type of high-workload situation.[6] Second, the flight crew were unable to *comprehend* the nature of the problem. This was in no small part to do with the presentation of contradictory and changing airspeed information. However, this was a known potential problem, and a relatively simple procedure existed to maintain safe flight during instances of unreliable airspeed indications. Third, the flight crew failed to project the results of their actions, such as increasing the nose up pitch attitude of the aircraft, which eventually resulted in the aerodynamic stall and loss of control. In all, there were issues with situation awareness at each of the three stages of Endsley's model.

The accident report highlights these issues in the harrowing transcript of the cockpit voice recorder (CVR), where the co-pilot in the left seat repeatedly states: 'what's happening? I don't know what's happening', and the final words of the co-pilot in the right seat are 'we're going to crash … this can't be true … but what's happening?'[5]

In a subsequent Human Factors analysis of the accident, the concept of the *system* as a whole suffering from deficiencies in situation awareness has been introduced.[7] This is an advanced concept but worth exploring in higher levels of situation awareness training sessions. This concept will be examined in more detail with respect to *distributed cognition* in the following section.

Situation Awareness: Core Enabling Knowledge

As with each of the non-technical skills, there is a core set of underlying theoretical knowledge that enables skilful performance. Achieving and maintaining high-level situation awareness can be best described as involving a coordinated set of integrated cognitive processes. Therefore, the underlying knowledge requirements span a number of different aspects of human cognition. In addition to the excellent overview provided in *Safety at the Sharp End*,[8] the following topics are important for an understanding of situation awareness.

The Basic Model of Situation Awareness

Naturally, an understanding of Mica Endsley's seminal model of situation awareness is the primary building block of knowledge required for enhancing this domain of non-technical skill. While this model is described in a number of human factors resources and textbooks, there are also a large number of research studies to draw from in many industry settings.

A Basic Introduction to Information Processing

A basic understanding of human information processing forms another important element of core enabling knowledge for situation awareness. The best source of this information would be an introductory psychology or human factors textbook. Key concepts here relate to the following topics:

- Our senses
- The sensory registers
- Perception
- Expectancy
- Short-term memory
- Working memory
- Top-down versus bottom-up processing

Shared Mental Models and Team Situation Awareness (TeamSA)

A more advanced set of knowledge relates to an understanding that situation awareness is not solely a feature of what is in the head of the individual operator. Rather, situation awareness is something that needs to be shared across a team. The concept of shared mental models is an important consideration and describes a congruent shared understanding of the situation, which is critical to safe and efficient performance. In particular, an understanding of the role of planning in establishing and maintaining a shared mental model is critical.[9,10]

Distributed Cognition and Distributed Situation Awareness

Similarly, situation awareness is supported by information processing both in the minds of individuals and in aspects of the technological systems we use. An understanding of distributed cognition is therefore important bedrock for effective situation awareness. This concept is introduced elegantly by Edwin Hutchins, who describes in the paper 'How a cockpit remembers its speeds' the phenomena of situation awareness being generated through information processing by the pilot and by the aircraft systems.[11] With respect to situation awareness, adopting a whole-of system-viewpoint has been termed *distributed situation awareness*,[12] and a body of research evidence is emerging to describe the features of distributed situation awareness in more detail.[13,14]

Cue Use and Cue Utilisation

An emerging body of research has focussed on the ways in which experts seek out critical information from the environment. There is an understanding that high levels of situation awareness are not simply created by being completely aware of everything that is happening, but rather, are constructed from the most important, and sometimes less salient, pieces of information. These pieces of information are described as *cues* and are seen as critical triggers for identifying changes in system state and to prompt appropriate decision-making. These specific cues have since formed the foundation of training programs that reinforce a set of these cues to be the focus of specific foci of attention to enhance situation awareness and decision-making.[15]

Inattentional Blindness

Inattentional blindness refers to the paradoxical phenomenon whereby something might be entirely within our visual frame, but we do not see it, as our attention is focussed elsewhere. The classic example of inattentional blindness is described in an experiment in which participants were asked to observe a video of two basketball teams, one wearing white uniforms and one wearing black. Participants were asked to count the number of passes of the basketball made by the white team. During the short video, a person in a gorilla suit walked from left to right across the frame. In this part of the experiment, fewer than 50%

of participants 'saw' the gorilla.[16] The context of inattentional blindness emphasises the role of purposeful focussed attention and the fact that we 'can look but fail to see' aspects of our environment.

Role of Situation Awareness in Error Detection

Another key area of understanding is the role that situation awareness plays in the detection and management of error. Maintaining an on-going model of the actual system state, as well as having an understanding of the planned system state, enables the process whereby errors can be detected when there is a mismatch between these two models.[17,18]

Influence of Distraction, Workload and Other Factors

The final element of core enabling knowledge relates to the range of factors that can negatively impact on situation awareness. Factors such as distraction, high workload, monotony and boredom, multiple competing tasks, highly salient features of the environment, and conflicting information are all aspects that can lead to degraded situation awareness. Further, the 'startle effect' has recently been the focus of research in the aviation domain, as it has been implicated in the fear- and stress-related degradation of situation awareness in a number of recent catastrophic accidents.[8] The use of specific case studies to introduce these concepts is often highly effective and engaging.

It is useful to distinguish between internal and external sources of distraction and misplaced attention. This is articulated elegantly in a study of risk factors associated with a signal passed at danger (SPAD) event, whereby train drivers are at risk both from environmental distractions (such as an event on the platform) and from internal distractions (such as pressure and anxiety associated with on-time running).[19]

Finally, extended time on task is another factor that has been demonstrated to be associated with degraded situation awareness, and the ability for operators to take short breaks is critical to enable a brief period of reduced concentration prior to re-establishing their level of attention and task focus.[20]

Situation Awareness Recovery

One novel concept relates to the specific skills required for recovery from degraded situation awareness caused by the factors described in the previous section. Recent research has highlighted that after an interruption or distraction, recovery of situation awareness is achieved by increasing scanning. Operators who are successful in situation awareness recovery also specifically guide this increased scanning towards cues that were being attended to prior to the distraction.[21]

Situation Awareness: Skill Development

General Targets for Skill Development

Situation awareness is not a single skill, but as discussed in this chapter, it is the product of a range of cognitive processes. Skill development in this domain therefore first involves training in strategies to enhance these cognitive processes.

First, skills in information acquisition are the very foundation of training for enhanced situation awareness. This is particularly the case because the majority of failures in situation awareness as a non-technical skill relate to the first stage, that of perception.[22] It is likely that the other stages of situation awareness, comprehension and projection, are heavily reliant on technical knowledge and expertise. However, if the first stage breaks down, this will have flow-on effects for the latter stages of situation awareness.

The training of specific skills in information acquisition is largely dependent on the sequence of work activity or the specific task at hand. In the aviation domain, pilots are trained in a set of systematic visual 'scan' sequences to ensure that all critical flight parameters are monitored. Different scan sequences are devised for different tasks or phases of flight and for individual operators depending on their role or position on the flight deck. These specific scan sequences are heuristics, rules of thumb that provide a robust structure for attention and information acquisition. However, in experts, these structures become more flexible, such that scanning and dwell time on system elements adapt to changing task demands.[23]

Second, skills in mental model development and maintenance form another foundation of situation awareness. Of these, planning and preparation are paramount, to develop a baseline mental model against which any deviations can be quickly identified. Moreover, as expectancy drives much of information acquisition in the dynamic environment, identifying the cues for critical pieces of information primes good situation awareness.[24] Therefore, training in a rigorous and systematic pre-job planning routine forms a critical part of training for enhanced situation awareness.

Individual versus Team Situation Awareness Training

There is a significant difference in the approach that needs to be adopted with respect to training for situation awareness at the individual level compared with the team level. For the individual, training focusses predominantly on the cognitive processes described above. However, at the team level, another set of processes is critical. Training at this level requires a focus on the skills associated with the development and maintenance of a shared mental model. This includes skills in planning, information exchange, resolution of conflicting understandings, workload and task delegation, and communication strategies more broadly.[25] Here, we see quite plainly the cross-over and interaction between aspects of non-technical skills.

Exposure to Factors That Degrade Situation Awareness

Beyond simply focussing on the development of skills in the underlying facets of situation awareness, it is also important to consider training in the management of factors that are known to lead to degraded situation awareness, as identified in the previous section. The use of data, such as from incident reports or investigations, to identify the underlying causes of degraded situation awareness has been identified as an important strategy for training needs analysis for situation awareness training programs.[26]

With respect to the 'startle effect' described previously, more frequent exposure to unexpected events and a focus on strategies to manage the negative effects of that startle have been suggested as highly beneficial.[27] It is likely that this training approach would generalise to other factors that can lead to degraded situation awareness, such as high workload conditions, stress and distractions.

Situation Awareness: Simulation-Based Training

Across a wide array of industries, simulation has become a critical component of training programs. With respect to situation awareness, simulation provides an environment in which powerful training curricula can be designed. As is the case with all forms of non-technical skill training, it has been demonstrated that skill development is significantly enhanced when immersive simulation-based training programs are adopted rather than more traditional classroom-based formats.[28] The following are some important considerations for simulation-based training for situation awareness.

Creating Rich Information Environments

As the underpinning skills associated with generating and maintaining situation awareness focus on information acquisition and assessment, it is critical that careful consideration is given to the design of an appropriately information-rich training scenario. While this does not necessitate ultra-high-fidelity simulation, it does demand consideration of all the relevant sources of information and critical cues associated with the training scenario.

At the most sophisticated level, a comprehensive cognitive work analysis (CWA) could be undertaken in the design of the training event.[29,30] The use of CWA enables the training designer to specify the various system 'inputs' for enhanced situation awareness and to identify the types of information available within the system and the ways in which agents exchange and otherwise interact with that information.

A less formal approach would be to list all the sources of information available to the training participants in generating high levels of situation awareness and to ensure that they are appropriately represented in the training scenario. This could simply be called *information fidelity* as a critical first consideration for situation awareness training.

Freeze Technique and Reflection/Debrief

Simulation-based training for situation awareness has two main benefits. First, it enables the scenario to be stopped at any point in time, such that aspects of situation awareness can be explored or assessed. This so-called 'freeze' technique was first suggested by Endsley as a technique for enabling the objective measurement of situation awareness by freezing a simulation and asking participants a series of probes relating to each of the three levels of situation awareness.[31] When augmented with the opportunity for critical reflection and debrief, this becomes a powerful training technique.

Recently, the benefits of this technique have been empirically demonstrated in diverse settings. One study demonstrated that this technique was associated with increased task performance, lower workload, and superior subjective and objective ratings of situation awareness compared with a control group who were only given traditional technical skills-based training.[32] This technique will be discussed in more detail in the following section on assessment of situation awareness.

Controlled Manipulation of Events to Degrade Situation Awareness and Demand Situation Awareness Recovery

Simulation-based training allows a much more controlled and safe introduction of factors that are known to degrade situation awareness. A simulator scenario can be crafted such that periods of high workload, distractions and surprise events can easily be scripted and participants must draw on each of the skills that underpin high levels of situation awareness.

Situation Awareness: On-the-Job Training

On-the-job training for situation awareness enables rehearsal, behavioural reinforcement and feedback through debriefing and coaching activities. The benefit of workplace-based training is that it presents a less predictable environment, where situation awareness must be dynamically maintained. Instructional processes therefore need to harness this unpredictability to enhance skill development.

Activation and Priming

Prior to workplace-based training for optimal situation awareness, prior knowledge and experience need to be activated.[33] This is achieved by refreshing the core enabling knowledge and asking the trainee to identify what aspects might be critical to the task at hand. For instance, this discussion might focus on the critical sources of information or on interpretation of those cues in creating a forward-looking view of the situation as it is unfolding.

This activation of prior knowledge can then transition into the identification of the core skills required to maintain optimal situation awareness. Also of considerable use is the trainee being asked to recall similar or analogous work to that occurring during the workplace training event.

Behavioural Modelling

The use of behavioural modelling, as another metacognitive training strategy, has been demonstrated to significantly enhance situation awareness. In a recent study, the situation awareness of learner drivers was found to be much better in a group who had been guided to monitor the driving behaviour of other drivers, identify the strategies that they use to maintain situation awareness, put those strategies into practice and reflect on their performance.[34,35]

This approach, which is aligned with the concept of the *reflective practitioner* discussed in Chapter 3, can facilitate the learning of effective cue utilisation and interpretation.

Situation Awareness: Pre-Training Briefing

Briefing situation awareness prior to training should focus on three major aspects. First, briefing should discuss the overall key performance outcomes: in this case, an optimal level of task-specific situation awareness. This sets the expectations of adequate performance and allows the general definitions of situation awareness to be revised and then contextualised for the specific training scenario.

Second, briefing should review the underlying enabling knowledge of situation awareness, as described previously. This review should highlight those aspects that are specifically relevant to the training scenario, such as distributed situation awareness.

Third, briefing should explore the specific cognitive skills and cognitive strategies that are to be drawn on in the scenario, as well as the key sources of information. If there is a specific heuristic to be deployed, such as a scanning technique, this should be mentally rehearsed in briefing prior to the training session.

AN EXAMPLE IN PRACTICE

In the metropolitan rail environment, a driver's mental model of a situation is created through the complex interplay of pre-existing knowledge of the route, monitoring of position along the route, monitoring of speed and other in-cab cues relating to train systems, monitoring of trackside infrastructure such as signals, monitoring the rail corridor for hazards, and identification of key cues for train-handling, such as acceleration or braking.

A simulator scenario has been developed to rehearse the key skills involved in creating and maintaining an accurate mental model, and identifying and responding to hazards such as signalling failures or trackside dangers in a timely manner. The instructor guide outlines a series of questions, which are to be used as points for facilitated discussion between instructor and trainee. These questions include:

- What are the main cues for acceleration and braking in this section of track?
- What are the main sources of distraction that need to be kept in mind?
- What critical information might I miss?
- When is it critical to be looking out, rather than heads down in-cab?

Situation Awareness: Assessment and Behavioural Markers

As discussed in Chapter 5, for a non-technical skill to be assessed, it first needs to be possible to identify, and second, it needs to be possible to measure. Although the domain of situation awareness is broad, much research has been undertaken with respect to the ways in which situation awareness can be observed and measured.[31] In turn, this research tells us much about how individuals' and team's situation awareness can be assessed.

The first consideration relates to our discussion of situation awareness being a multi-faceted domain of non-technical performance, made up of a number of underlying cognitive skills at the individual level, a set of interpersonal skills at the team level, and also the interaction between the human and non-human components of a system at the distributed level. To this end, when we refer to assessing situation awareness, a decision needs to be made on whether we are assessing the emergent outcome of these underlying skills, that is to say the *degree* of situation awareness, or whether we are assessing at a more procedural level the performance in each of a set of underlying skills. Neither of these approaches is right or wrong; rather, they need to be clearly considered and addressed in the design of any form of situation awareness assessment.

The research literature relating to the measurement of situation awareness provides us with some significant insights into how we can assess situation awareness. One of the first, and most popular, techniques for measuring situation awareness is the Situation Awareness Global Assessment Technique (SAGAT).[31] The SAGAT approach is one that is used when the training situation can be temporarily 'frozen' such that participants can be asked a series of queries relating to any aspects of the situation that the participant would normally be attending to and processing. These queries are designed specifically with the training scenario in mind.

Other popular techniques for the assessment of situation awareness are observational in nature and draw on expert observers rating specific behavioural indicators of situation awareness. The key generic aspects of such 'behavioural marker systems' have been discussed in detail in Chapter 5, so we will simply discuss some specific examples here.

One typical behavioural marker system for situation awareness specifies three elements under the category of situation awareness. As described in Table 7.1, each of the elements maps onto the one of the three levels of situation awareness described in Endsley's model. This general framework for specifying behavioural markers associated with situation awareness can be found in the anaesthetists' non-technical skills (ANTS) framework,[36] the Non-Technical Skills for Surgeons (NOTSS) framework[37] and the Well-Operations Crew Resource Management (WOCRM) framework,[38] among others.

Table 7.1 Typical Elements within the Domain of Situation Awareness

CATEGORY	ELEMENT
Situation awareness	Gathering information Understanding information and risk status Anticipating future state/developments

Table 7.2 Example Behavioural Markers for Situation Awareness Element 'Gathering Information' in the ANTS Framework

ELEMENT	BEHAVIOURAL MARKERS FOR GOOD PRACTICE
Gathering information	• Obtains and documents patient information pre-operatively • Conducts frequent scan of the environment • Collects information from team to identify problem • Watches surgical procedure, verifying status when required • Cross-checks information to increase reliability
	BEHAVIOURAL MARKERS FOR POOR PRACTICE
	• Reduces level of monitoring because of distractions • Responds to individual cues without confirmation • Does not alter physical layout of workspace to improve data visibility • Does not ask questions to orient self to situation during handover

Source: Fletcher, G. et al., *British Journal of Anaesthesia*, *90*(5), 580–588, 2003.

Specific behavioural markers can then be developed for each of the elements as they relate to the specific job-roles, tasks or scenarios. As discussed in Chapter 5, there are significant benefits associated with developing detailed sets of behavioural markers that are specific to the individual training scenario. This provides the instructors with excellent guidance on how the specific aspects of situation awareness, and the underlying skills involved, are important to achieving safe and efficient operations at the task level. However, the majority of existing behavioural marker systems provide guidance only for good and poor behaviours at the generic global level, as per Table 7.2.

Situation Awareness: Debriefing and Coaching

Debriefing situation awareness, again, can focus on either the top-level outcome, the overall degree of situation awareness held by an individual or team, or the specific skills and strategies used to achieve this outcome. Indeed, a comprehensive debrief will unpack and provide coaching around both of these aspects.

Unlike other forms of non-technical skills, the use of video in debriefing probably has less value when it comes to individual assessment of situation awareness. So much of situation awareness at the individual level involves internal cognitive processes that video might only serve as a memory prompt to what the participant might have been focussing their attention on or thinking about at that point in time. Trigger questions for debriefing individual situation awareness might include the following:

- What was the first thing that alerted you to the situation?
- What pieces of information were critical at this stage of the situation?
- Was there any information you felt was missing?
- What was your understanding of the current situation at this point in time?
- Did you consider that something else might have been more likely?
- At this point in time, what was your thinking in relation to where the situation was heading?
- Was there any point in time that you felt you didn't understand what was happening?

Coaching individual situation awareness should focus on each of the key cognitive skills that are used to generate and maintain situation awareness.

When it comes to debriefing team situation awareness, video is potentially highly useful. A video can be paused at a critical point, such as after a period of team information exchange, and participants can be prompted to discuss what was their understanding of the situation at that point in time. In addition to asking individual participants questions listed in the preceding paragraph, trigger questions for a video-assisted debrief of team situation awareness might include:

- How was critical information shared among the team?
- Did you consider that you maintained a shared understanding of the situation as it unfolded?
- What parts of the system (people or technology) held the critical pieces of information?
- How did you go about identifying the situation, and what were the potential risks associated with events as they unfolded?
- What could you have done better to ensure a shared mental model?

Situation Awareness: Key Resources

Flin, R. H., O'Connor, P., and Crichton, M. (2008). *Safety at the Sharp End: A Guide to Non-Technical Skills*. Aldershot, UK: Ashgate Publishing.

Endsley, M. R. (1995a). Measurement of situation awareness in dynamic systems. *Human Factors*, *37*(1), 65–84.

Endsley, M. R. (1995b). Toward a theory of situation awareness in dynamic systems. *Human Factors*, *37*(1), 32–64.

Case Studies

Three Mile Island: In the early hours of 28 March 1979, a series of failures led to the loss of coolant to one of the reactors of the Three Mile Island nuclear energy plant in Harrisburg, Pennsylvania. The loss of coolant, and subsequent overheating of the nuclear core, led to a very nearly catastrophic accident. As part of the accident sequence, the control room operators failed to notice in a timely manner that critical valves had remained closed after maintenance had been performed in the days prior. Furthermore, the design of displays in the control room allowed the control room operators to assume other valves were closed, when in fact they were open. These failures of situation awareness led to the loss of coolant to the reactor. This case study involves issues relating to situation awareness that have their origins in system design, display locations and formats, and inaccurate mental models of the control room operators.

Elaine Bromiley: This case study involves a patient who tragically died while undergoing a routine operation. After commencing the induction of anaesthesia, the anaesthetist had difficulty maintaining oxygen levels and was unable to maintain an effective airway with several different sizes of laryngeal mask. The anaesthetist then attempted to intubate Mrs Bromiley, unsuccessfully. This situation escalated over the next few minutes to a situation known as 'can't intubate, can't ventilate', where the accepted protocol is to access the airway through a tracheotomy (incision into the windpipe). However, for the next 30 minutes, the anaesthetist and other colleagues continued with attempts to intubate. After over 30 minutes with very low blood oxygen levels, the procedure was abandoned, but Mrs Bromiley never regained consciousness.[39] The case can be explored with respect to failures of DSA, with several good videos available online to support the case.

References

1. Endsley, M. R. (1988). Design and evaluation for situation awareness enhancement. In *Proceedings of the Human Factors and Ergonomics Society Annual Meeting* (Vol. *32*, pp. 97–101). Santa Monica, CA: Sage.
2. Endsley, M. R. (1995). Toward a theory of situation awareness in dynamic systems. *Human Factors: The Journal of the Human Factors and Ergonomics Society*, *37*(1), 32–64.
3. Dekker, S., and Hollnagel, E. (2004). Human factors and folk models. *Cognition, Technology & Work*, *6*(2), 79–86.
4. Dekker, S. W., Hummerdal, D. H., and Smith, K. (2010). Situation awareness: Some remaining questions. *Theoretical Issues in Ergonomics Science*, *11*(1–2), 131–135.
5. Bureau d'Enquêtes et d'Analyses. (2012). Final report on the accident on 1st June 2009 to the Airbus A330-203 registered F-GZCP operated by Air France flight AF 447 Rio de Janeiro–Paris. Paris: BEA.
6. Colavita, F. B. (1974). Human sensory dominance. *Perception & Psychophysics*, *16*(2), 409–412.
7. Salmon, P. M., Walker, G. H., and Stanton, N. A. (2016). Pilot error versus sociotechnical systems failure: A distributed situation awareness analysis of Air France 447. *Theoretical Issues in Ergonomics Science*, *17*(1), 64–79.
8. Flin, R. H., O'Connor, P., and Crichton, M. (2008). *Safety at the Sharp End: A Guide to Non-Technical Skills*. Aldershot, UK: Ashgate Publishing.
9. Converse, S. A., Cannon-Bowers, J. A., and Salas, E. (1991). Team member shared mental models: A theory and some methodological issues. In *Proceedings of the Human Factors Society 35th Annual Meeting* (pp. 1417–1421). Santa Monica, CA: Human Factors and Ergonomics Society.
10. Stout, R. J., Cannon-Bowers, J. A., Salas, E., and Milanovich, D. M. (1999). Planning, shared mental models, and coordinated performance: An empirical link is established. *Human Factors*, *41*(1), 61–71.
11. Hutchins, E. (1995). How a cockpit remembers its speeds. *Cognitive Science*, *19*, 265–288.
12. Stanton, N. A., Stewart, R., Harris, D., Houghton, R. J., Baber, C., McMaster, R., Salmon, P., Hoyle, G., Walker, G., and Young, M. S. (2006). Distributed situation awareness in dynamic systems: Theoretical development and application of an ergonomics methodology. *Ergonomics*, *49*(12–13), 1288–1311.
13. Fioratou, E., Flin, R., Glavin, R., and Patey, R. (2010). Beyond monitoring: Distributed situation awareness in anaesthesia. *British Journal of Anaesthesia*, *105*(1), 83–90.

14. Schulz, C. M., Endsley, M. R., Kochs, E. F., Gelb, A. W., and Wagner, K. J. (2013). Situation awareness in anesthesia: Concept and research. *The Journal of the American Society of Anesthesiologists*, *118*(3), 729–742.
15. Wiggins, M., and O'Hare, D. (2003). Weatherwise: Evaluation of a cue-based training approach for the recognition of deteriorating weather conditions during flight. *Human Factors: The Journal of the Human Factors and Ergonomics Society*, *45*(2), 337–345.
16. Simons, D. J., and Chabris, C. F. (1999). Gorillas in our midst: Sustained inattentional blindness for dynamic events. *Perception*, *28*(9), 1059–1074.
17. Thomas, M. J. W., and Petrilli, R. M. (2004). Error management training: Defining best practice report 2004/0050. Canberra, Australia: Australian Transport Safety Bureau.
18. Thomas, M. J. W., and Petrilli, R. M. (2007). Error management training: Identification of core cognitive and interpersonal skill dimensions. In J. M. Anca (Ed.), *Multimodal Safety Management and Human Factors* (pp. 169–178). Aldershot, UK: Ashgate Publishing.
19. Naweed, A. (2013). Psychological factors for driver distraction and inattention in the Australian and New Zealand rail industry. *Accident Analysis & Prevention*, *60*, 193–204.
20. Sneddon, A., Mearns, K., and Flin, R. (2006). Situation awareness and safety in offshore drill crews. *Cognition, Technology & Work*, *8*(4), 255–267.
21. Kaber, D. B., Riley, J. M., Endsley, M. R., Sheik-Nainar, M., Zhang, T., and Lampton, D. R. (2013). Measuring situation awareness in virtual environment-based training. *Military Psychology*, *25*(4), 330.
22. Jones, D. G., and Endsley, M. R. (1996). Sources of situation awareness errors in aviation. *Aviation, Space, and Environmental Medicine*, *67*(6), 507–512.
23. Bellenkes, A. H., Wickens, C. D., and Kramer, A. F. (1997). Visual scanning and pilot expertise: The role of attentional flexibility and mental model development. *Aviation, Space, and Environmental Medicine*, *68*(7), 569–579.
24. Endsley, M. R., and Robertson, M. M. (2000). Training for situation awareness in individuals and teams. In M. R. Endsley and D. J. Garland (Eds.), *Situation Awareness Analysis and Measurement* (pp. 349–366). Mahwah, NJ: Lawrence Erlbaum Associates.
25. Salas, E., Prince, C., Baker, D. P., and Shrestha, L. (1995). Situation awareness in team performance: Implications for measurement and training. *Human Factors*, *37*(1), 123–136.
26. Endsley, M. R., and Garland, D. J. (2000). Pilot situation awareness training in general aviation. In *Proceedings of the Human Factors and Ergonomics Society Annual Meeting* (Vol. *44*, pp. 357–360). Santa Monica, CA: Sage.
27. Martin, W. L., Murray, P. S., Bates, P. R., and Lee, P. S. (2015). Fear-potentiated startle: A review from an aviation perspective. *The International Journal of Aviation Psychology*, *25*(2), 97–107.

28. Nazir, S., Sorensen, L. J., Øvergård, K. I., and Manca, D. (2015). Impact of training methods on distributed situation awareness of industrial operators. *Safety Science*, *73*, 136–145.
29. Naikar, N., and Sanderson, P. (1999). Work domain analysis for training-system definition and acquisition. *The International Journal of Aviation Psychology*, *9*(3), 271–290.
30. Naikar, N., and Saunders, A. (2003). Crossing the boundaries of safe operation: An approach for training technical skills in error management. *Cognition, Technology and Work*, *5*, 171–180.
31. Endsley, M. R. (1995). Measurement of situation awareness in dynamic systems. *Human Factors: The Journal of the Human Factors and Ergonomics Society*, *37*(1), 65–84.
32. Saus, E.-R., Johnsen, B. H., Eid, J., Riisem, P. K., Andersen, R., and Thayer, J. F. (2006). The effect of brief situational awareness training in a police shooting simulator: An experimental study. *Military Psychology*, *18(S)*, S3.
33. Merrill, M. D. (2002). First principles of instruction. *Educational Technology Research and Development*, *50*(3), 43–59.
34. Kim, B., Park, H., and Baek, Y. (2009). Not just fun, but serious strategies: Using meta-cognitive strategies in game-based learning. *Computers & Education*, *52*(4), 800–810.
35. Soliman, A. M., and Mathna, E. K. (2009). Metacognitive strategy training improves driving situation awareness. *Social Behavior & Personality: An International Journal*, *37*(9), 1161–1170.
36. Fletcher, G., Flin, R., McGeorge, P., Glavin, R. J., Maran, N. J., and Patey, R. (2003). Anaesthetists' non-technical skills (ANTS): Evaluation of a behavioural marker system. *British Journal of Anaesthesia*, *90*(5), 580–588.
37. Yule, S., Flin, R., Maran, N., Rowley, D., Youngson, G., and Paterson-Brown, S. (2008). Surgeons' non-technical skills in the operating room: Reliability testing of the NOTSS behavior rating system. *World Journal of Surgery*, *32*(4), 548–556.
38. Flin, R., Wilkinson, J., and Agnew, C. (2014). *Crew Resource Management for Well Operations Teams*. London, UK: International Association of Oil & Gas Producers.
39. Bromiley, M. (2008). Have you ever made a mistake? *A Patient Liaison Group Debate. The Royal College of Anaesthetists Bulletin*, *48*, 2442–2445.

8

Training and Assessing Decision-Making

Decision-Making: A Primer

In any high-risk industry, the real operating environment is often nothing like the neat and constrained world of the standard operating procedures. Operators are frequently faced with non-normal and sometimes unexpected situations to which they must effectively respond to maintain safe and efficient operations. From the identification that something is not as planned, through diagnosis of the situation, through to making a suitable choice of action is collectively known as *decision-making*.

It is one of the unique features of being human, and one of our greatest strengths, that we are able both to reason inductively and to make judgements on the most appropriate course of action. This is a primary feature of our human intelligence and something that the technological systems we operate almost completely lack.

The way in which we undertake the decision-making process is highly dynamic and task specific. There are many forms of decision-making, and just as many theoretical frameworks have been proposed to explain how we make complex decisions.

The classical model of decision-making, sometimes referred to as *rational* decision-making, describes the conscious and effortful process of identifying the problem, generating potential solutions, weighing up each of these options, choosing the most appropriate option, putting it in place and evaluating whether it achieves the desired outcome. This effortful, rational approach is sometimes described in terms of the basic information-processing model introduced in the previous chapter. Table 8.1 describes the basic stages of decision-making and their relationship with elements of information processing.[1]

Table 8.1 Decision-Making Described in Terms of Information-Processing Stages

STAGE OF DECISION-MAKING	DESCRIPTION
Cue reception and integration	The first stage involves the initial information input for the decision-making process, whereby cues in the environment are received and integrated. This maps onto the information-processing elements of attention and perception.
Hypothesis generation	The second stage involves interpreting those cues and forming a complete understanding of the situation. This maps onto the sense-making stages of information processing, bringing together top-down and bottom-up processing of the environmental cues and generating hypotheses about the situation at hand.
Hypothesis evaluation	The third stage of decision-making involves the evaluation of the hypotheses generated in the previous stage. Again, this is an effortful cognitive process drawing on pre-existing knowledge and experience.
Generating and selecting actions	The final stage of decision-making from an information-processing perspective involves generating actions in response to the situation. These actions are then evaluated with respect to expected and optimal outcomes.

Source: Wickens, C.D. et al., *An Introduction to Human Factors Engineering*, New York, NY, Addison Wesley Longman, 1998.

The messy nature of real-world operations in high-risk industries presents some challenges to the rational and effortful, and thus time-consuming, classical approach to decision-making. The complex context of high-risk industries is one where decision-making must occur within the constraints of limited information and finite resources and time.[2,3] These constraints place limitations on the degree to which all decisions can be made, given the resources available.

This concept of 'bounded rationality', where the decision space is highly constrained, gave rise to the concept of System 1 and System 2 decision-making. This differentiation proposes that due to the challenges of bounded rationality, many decisions are made in a more intuitive or *naturalistic* form of decision-making (System 1) than a formal *rational* manner (System 2).[4]

In response to the challenge of bounded rationality, Gary Klein developed a model of System 1 decision-making, which he named *recognition-primed decision-making*.[5] Put simply, the model describes how experts make decisions, drawing on their knowledge and experience to produce a safe and effective outcome within the constraints of bounded rationality.

As a form of naturalistic decision-making, recognition-primed decision-making sees the central component of decision-making as the matching of situations to pre-rehearsed scenarios that already have mapped out appropriate courses of action.[6] Once a situation is matched against a pre-rehearsed scenario, required actions are implemented, supported by four simple processes: (1) setting *plausible goals* for responding to the situation; (2) creating *expectancies* that describe the anticipated characteristics of the situation against which the unfolding events can be evaluated; (3) identifying the *critical cues* needed to verify that the diagnosis of the situation was correct in the first place; and (4) defining the obvious first *course of action* in terms of an optimal response strategy.[7]

The recognition-primed decision-making model then describes more complex processes involved in mental simulation of the action to support evaluation of the proposed actions prior to implementing them, or further diagnostic processes in the event that the situation does not exactly match a pre-rehearsed scenario. These aspects of the recognition-primed decision-making model highlight how experts make decisions and also build more complex schemata of pre-rehearsed solutions to unique problems as they are experienced in actual or simulated situations.

Decision-Making: Exemplar Case Study

On 3 July 1988, the USS *Vincennes* shot down Iran Air Flight 655 over the Persian Gulf, killing 290 civilians, in a tragic example of flawed decision-making. At the time of the event, the USS *Vincennes* was involved in active engagement with Iranian forces in the Persian Gulf as part of an operation to prevent Iranian forces from disrupting commercial shipping in the Persian Gulf.

On the morning of 3 July, the USS *Vincennes* was providing support to the USS *Elmer Montgomery*, which at the time was surrounded by 13 Iranian gunboats. Part of this support involved the dispatch of one of USS *Vincennes*' patrol helicopters to the vicinity of the activity, and on arrival, the helicopter was fired on by one of the Iranian vessels. On approach to the USS *Elmer Montgomery*, these Iranian gunboats closed in on, and opened fire on, the USS *Vincennes*. The USS *Vincennes* was then involved in surface engagement as well as attempting to provide protection to the helicopter.

While this active engagement with Iranian vessels was occurring, Iran Air Flight 655 took off from Bandar Abbas airport, which was a joint military and civilian airfield. The flight was a routine scheduled commercial passenger flight and was routed along a declared civilian airway. The commercial airliner was broadcasting a civilian transponder code and was in active communication with Bandar Abbas air traffic control. Iran Air Flight 655 was cleared to an altitude of 14,000 feet and was climbing to this altitude throughout the seven minutes from take-off to being shot down by the USS *Vincennes*. The whole scenario unfolded in just over three minutes, from the first identification of the aircraft to the firing of the missile that brought the airbus down.

At 0947 local time, immediately after take-off, Iran Air Flight 655 was detected on radar by the USS *Vincennes* and was assigned an identifier of TN4131. The aircraft was seen to be on a direct course towards USS *Vincennes* and was closing in rapidly. At 0949, USS *Vincennes* began warnings to the aircraft on military and international distress channels. One minute later, at 0950, the USS *Vincennes* heard on US military radio channels a report of F-14 Iranian combat aircraft activity. During this time, surface combat was continuing, and the USS Vincennes suffered a fouled gun mount, which meant that the vessel had to manoeuvre at speed. Also, the USS *Vincennes* was monitoring an Iranian P-3 military aircraft in the vicinity that was flying a profile associated with providing information to attack aircraft. Warnings continued to the aircraft, but after no response, or any change in course, the commanding officer of the USS *Vincennes* made a decision to engage, and two missiles were launched, which intercepted Iran Air Flight 655 at a range of eight NM from USS *Vincennes* at an altitude of 13,500 feet.[8]

In this seven minute period of time, a range of factors influenced the erroneous decision to fire on the Iranian airliner. First, the decision was significantly constrained by time limitations, with only a matter of minutes from the detection of the target to the requirement to engage prior to the USS *Vincennes* being left largely unprotected against attack from the aircraft. Second, the decision was made within the context of a stressful situation, in which multiple competing demands, and multiple threats, further constrained the resources available to decision-making. Further, the commanding officer and team aboard the USS *Vincennes* made the decision to fire on the aircraft in a complex context that resulted in the *expectation* of hostile engagement from the Iranian air force. Against the backdrop of several weeks of increasing hostility and the surface combat at the time of the event, a frame of attack primed the decision. This is highlighted in the terminology used by the US Navy that the target was 'unidentified assumed hostile'.

This case study highlights several critical aspects of the domain of decision-making in the context of skill development and training. First, it highlights that in high-risk industries, decision-making is performed in a highly constrained environment, where decisions need to be taken in a time-critical fashion with often quite limited information available. Second, it highlights that in high-risk industries, decision-making is frequently a team-based operation, which raises a number of complexities including information transfer, shared mental models, and more subtle influences on the decision-making process at the hands of social psychology. Third, it highlights the influence of contextual factors such as time pressure and stress on decision-making in high-risk industries. Finally, the case study introduces the influence of heuristics and cognitive biases, for instance mindset, expectation and confirmation bias in the decision-making process. Each of these areas informs critical considerations for the design of non-technical skills training programs for the decision-making domain.

Decision-Making: Core Enabling Knowledge

The decision-making of operators in high-risk industries has been the focus of considerable research over the past several decades. Accordingly, we now have a significant body of knowledge on which we can build our non-technical skills training programs. The following outlines the areas of core enabling knowledge with respect to decision-making.

Classical Model of Decision-Making

The obvious place to begin a training program is to reinforce the basic precepts of the classical model of decision-making. There is a lovely axiom used in many industries that refers to decisions that are '*fast, decisive … but wrong!*' To avoid making errors of judgement in the heat of the moment, the classical model of decision-making is often reinforced in training, and mnemonics are used for a methodical and structured approach to decision-making. An example of such a mnemonic is provided in the skills development section of this chapter.

Throughout our lives, we make personal high-stakes decisions using the rational approach to decision-making, where we weigh up the pros and cons of a choice such as purchasing a house. An understanding of the classical model of decision-making can be augmented by introducing the concepts of inductive (from specifics to a theory) and deductive reasoning (from the general theory to specifics). Identifying examples of both forms of reasoning at play in high-risk decision-making is a part of core knowledge development.

Bounded Rationality in Context

Another component of core enabling knowledge is to develop an understanding of bounded rationality in the context of the specific work environment. The constraints on slow and methodical decision-making are typically unique to the work environment. The information, resource and time constraints placed on a site supervisor managing the breakdown of a coolant system are significantly different from those facing a doctor making a diagnostic decision in the emergency room. Sound knowledge of the constraints found in the high-risk work environment enables tactical changes to the decision-making process to be made by operators in the dynamic work environment.

Recognition-Primed Decision-Making and Fast and Frugal Approaches

A detailed understanding of how experts make decisions in dynamic environments is also critical enabling knowledge. The first element of core knowledge here relates to developing and understanding System 1 and System 2 decision-making and the differences between the two.

A training program may then decide which of the many models of System 1 decision-making will be explored in detail. While traditionally Klein's recognition-primed model[5,7] has been the most common of these to be explored in traditional crew resource management programs, more recently there have been a number of alternate models that offer useful new perspectives on System 1 decision-making.

System 1 decision-making has also been the subject of a number of popular texts, such as Malcolm Gladwell's *Blink*,[9] Daniel Kahneman's *Thinking, Fast and Slow*[10] and others. These texts provide engaging examples to convey the core concepts associated with these forms of decision-making.

Heuristics and Cognitive Biases

Heuristics and cognitive biases refer to the mental short cuts and 'rules of thumb' we use in our everyday decision-making. Most of the time, these short cuts are of significant benefit. They serve an important purpose in enabling complex decision-making, and from an evolutionary perspective, they have enabled us to make complex decisions without the continual investment of energy in detailed rational thought. Heuristics and biases are sometimes referred to as *cognitive dispositions to respond*, as they describe 'hard wired' features of our decision-making.[11] Some of the more common heuristics and biases that influence decision-making are summarised in Table 8.2.

Table 8.2 Heuristics and Biases That Influence Decision-Making

HEURISTIC/BIAS	DESCRIPTION
Availability heuristic	We are more likely to choose something that readily comes to mind, such as the more common or easily recognised option. This is especially the case if we have seen a similar situation recently.
Confirmation bias	We have a tendency to seek information that will confirm our initial thinking rather than see information that might disprove our hypothesis.
Commission bias	We are more inclined to take action rather than sit back and monitor the situation. This reflects the urge just to *do something*.
Search satisficing	We are inclined to stop seeking information once we have identified a likely problem. However, further examination may reveal critical information relevant to the decision-making process.
Decision momentum	Once a reasonable alternative has been identified, it is harder and harder to choose another option.
Overconfidence bias	A tendency to think our knowledge and skills are better than they actually are in reality. For example, if you ask a room of professionals to identify whether they think they are below the group average, far fewer than 50% will do so.
Optimism bias	We are more inclined to over-estimate the likelihood of positive outcomes than of more unpleasant alternatives. This is encapsulated nicely in the Australian idiom 'She'll be right, mate.'

Source: Croskerry, P., *Academic Emergency Medicine, 9*(11), 1184–1204, 2002.

Team Decision Processes

In many high-risk work environments, decision-making is not a solo endeavour. Frequently, small and sometimes larger teams make decisions that come with significant safety implications. In this context, we can define team decision-making in terms of two or more individuals with their own domain expertise coming together to make operational decisions. In high-risk industries, teams are often made up of individuals with quite different specialist knowledge and skills, working towards a common goal. The field of Human Factors has highlighted a number of ways in which team decision processes can be optimised and the ways in which team decision-making can be compromised.

Optimising team decision-making involves ensuring that all relevant information is available to all team members and that the specialist expertise of individuals is drawn on equally. Further, two critical coordination functions have been identified as important to all forms of teamwork: *shared mental models* and *closed loop communication.*[12] As with many aspects of non-technical skills, it is evident here that there are interdependences, and here, aspects of situation awareness and communication are critical to effective decision-making in the team environment. In many respects, optimising team decision-making also involves conscious management of a number of potential issues that may compromise the process and lead to a sub-optimal outcome.

One of the most famous ways in which team decision-making can be compromised is described through social psychology in terms of *group-think*. In the team situation, the natural desire to achieve consensus often interferes with the required processes for individual analysis of information and providing expert opinion to the team. Group-think refers to situations where individuals do not want to 'rock the boat' and in turn, do not question the direction in which the team is heading in terms of a decision.[13]

Another factor that can compromise team decision-making is when the team is neither co-located nor necessarily working towards one single common goal. We refer to these types of teams as undertaking *distributed decision-making*.

Distributed Decision-Making

Sometimes, decisions must be made by groups of people who might be distributed in locality. Further, these distributed decisions often need to be made by people with very different job-roles and competing priorities, and who may not even have a completely common set of goals. Take, for example, the coordinated decision-making of air traffic controllers and flight crew in a situation that demands a diversion to the nearest airport. The flight crew have at their disposal a set of information and specialist expertise relating to their aircraft and the technical issues and associated constraints that it may be facing at the time. Their goal is to work within those constraints to land at the nearest suitable airport. The air traffic controller, on the other hand, is limited in their understanding of the situation facing the aircraft, but has a broader understanding of the other traffic in the vicinity of the stricken aircraft and the nearest airports. Their goal is to ensure the safety of all aeroplanes in the vicinity while facilitating the diversion of the stricken aircraft. Together, with different mental models of the situation and with slightly different goals, the flight crew and air traffic control must make a series of optimal decisions to enable landing at the nearest suitable airport.

This type of distributed decision-making presents an additional set of constraints and faces several additional modes of potential failure compared with individual or team decision-making.

Resent research has highlighted a number of ways in which this form of decision-making can breakdown. First, it can suffer from *operational disconnect*, where the actions or plans of one party do not align with the expectations of the other party; for instance, where the flight crew commence immediate descent due to their aircraft system malfunction, but the air traffic controller expects them to maintain altitude until the aircraft is cleared to descend. Second, distributed decision-making can break down due to *informational disconnects*, where a difference in information occurs between elements of the distributed team; for instance, where the flight crew are unaware of unidentified traffic in their descent path. Finally, distributed decision-making can break down due to *evaluative disconnects*, where the evaluation of information differs between parties; for instance, where air traffic control do not appreciate the gravity of the emergency on the aircraft.[14] An understanding of these patterns of breakdown is core enabling knowledge for enhancing the decision-making skills of operators.

Decision-Making: Skill Development

While decision-making is often seen as a skill that develops as a component of expertise, it is still possible to develop and enhance decision-making skills through directed training. Unlike other domains of non-technical skills, however, one of the primary premises of decision-making training is that it is very difficult to achieve effective training outside the normal context in which decisions would normally be made.[15] That is to say, skill development in decision-making is difficult to achieve simply through the practice of generic abstracted cognitive tasks. Rather, there is a requirement to develop those skills within the context of domain knowledge and technical expertise.

Given this constraint, there is a need to focus on the use of simulation-based and on-the-job training programs, and also for developing training interventions that draw on highly realistic work-based scenarios.

Basic Foundational Skills: Situation Awareness, Risk Assessment and Evaluation

The first consideration in the development of enhanced decision-making skills is that the domain of situation awareness is highly intertwined with effective decision-making performance, and to a large degree, it can be said that situation awareness is a pre-requisite for the decision-making process. In reality, in the dynamic world of high-risk industries, such a linear approach is overly simplistic. However, in terms of training program design, it makes sense to focus on enhancing situation awareness skills prior to focussing on decision-making.

In high-risk industries, an understanding of risk and the ability to undertake effective risk assessment is another skill that forms a foundation of effective decision-making. Being able to identify the potential sources of risk associated with different options in a decision-making situation is necessary to ensure that safety is not compromised.

Finally, the process of evaluating the decision involves on-going monitoring of the situation and comparing the results of the decision with the initial desired outcomes. Again, aspects of situation awareness and risk assessment are both parts of this process.

Systematic Decision-Making Processes

Once the core enabling knowledge and the basic underpinning skills are in place, the most common approach to the development of decision-making skills involves the training of systematic structured processes that guide decision-making. The highly rational approach to decision-making set out in the classical model has been embedded in several mnemonics designed to guide operators in undertaking an effective decision-making process. These mnemonics are designed to lead both individuals and teams through a step-by-step process such that all critical information has been assessed and all potential options evaluated. One of these mnemonics is referred to as *A-GRADE*, and has been used in the aviation domain to facilitate flight crew decision-making, as summarised in Table 8.3.[16]

De-Biasing

As with all non-technical skills, a critical component of training involves the development of strategies to avoid or overcome things

Table 8.3 A-GRADE Model of Decision-Making

MNEMONIC	DESCRIPTION
A: Aviate	The first task is to *aviate*, ensuring that positive control of the aircraft is maintained at all times.
G: Gather	The second step is to *gather* all relevant information using all the resources available.
R: Review	The next step is to *review* the information you have gathered and identify whether there are any critical pieces of information missing.
A: Analyse	Next is to *analyse* all the options available to you, based on the information you have gathered and reviewed.
D: Decide	Once these steps have been completed, you can *decide* on a suitable course of action and put it into play.
E: Evaluate	Finally, you need to *evaluate* the decision, to ensure that the outcomes are as intended. If not, then the process needs to start again until you find an optimal solution.

Source: Edkins, G.D., *Flight Safety Australia*, September, 31–33, 2000.

that might negatively affect our performance. With respect to decision-making, this involves developing skills in avoiding the negative impacts of heuristics and cognitive biases. Recent research has highlighted the ability for strategies to be used to effectively 'de-bias' our decision-making.[17,18]

The principal method used to avoid the negative consequences of cognitive biases in decision-making is *metacognition*, which involves stepping away from the situation for a moment to examine and reflect on the thinking process.[18] This process involves first identifying situations in which a bias might influence the decision being made, and then using a cognitive forcing strategy to step back and purposefully examine the decision process for the presence of that bias. This is indeed an advanced decision-making technique, but the history of accidents provides a number of examples where disaster might have been averted if the crew had stepped back for a moment and examined their decision-making. For instance, take Air France Flight AF447, where the crew did not diagnose the aerodynamic stall of the aircraft; British Midland Flight 92, which crashed at Kegworth after the crew shut down the wrong engine; and the case of Elaine Bromiley, where anaesthetists persevered in trying to intubate Mrs Bromiley for over 20 minutes, resulting in her death from a prolonged period of lack of oxygen to the brain. Each of these tragic events might have been prevented by a brief metacognitive reflection on the decision-making process.

Analogical Reasoning

The development of decision-making skills relating to ill-defined problems where complex reasoning is required necessitates the development of analogical reasoning skills. One way to develop these types of skills is the case study techniques introduced in Chapter 4. Working on ill-defined case-based problems develops skills in case-based reasoning, which is a form of analogical reasoning whereby, when a novel problem is encountered, decision-making draws on previous problems that exhibit similarities to the current situation.[19]

The role of case-based reasoning in developing decision-making skills may have particular relevance to how so-called 'black swan' events are managed in high-risk industries. A black swan event is defined as a highly unpredictable event in which a situation occurs that was not considered possible in the design of the system.[20]

The development of these skills, which are more aligned with System 2 thinking, is perhaps the one exception to the general need for decision-making training to be situated in authentic learning environments such as the simulator or on-the-job training. The use of case study techniques has been demonstrated to achieve high levels of transfer of training,[19] most likely due to the fact that the problems are novel in nature and require detailed reasoning to solve.

Team-Based Training

Due to the prevalence of team-based work in high-risk work environments, the development of skills in decision-making should also include skills in optimising team performance and avoiding the factors that can compromise team-based decision-making, explored earlier in this chapter.

Once a basic understanding of the opportunities and constraints of team-based decision-making has been established, training in real-world teams can then take place and focus on the key aspects of developing shared mental models and effective communication in supporting the decision-making process.

Decision-Making: Simulation-Based Training

Due to the need for decision-making training to focus on solving real-world problems, and the need for such training to draw on technical expertise, the simulator provides an excellent environment for the development of individual and team-based decision-making skills.

Critical Decision Events

One of the most effective ways of developing training interventions in the simulator for decision-making is to develop scenarios in which the original plan becomes marginal and a decision needs to be made either to continue with the original plan or to generate possible alternative courses of action.

The creation of these scenarios can draw heavily on the evidence base used in training needs analysis, such as the data captured within an organisation's safety management system from normal operations monitoring or from incident investigations, as described in Chapter 6.

Building Expertise and a Repository of Experience

Research undertaken into the nature of decisions made by experts, such as the recognition-primed decision-making model described previously, highlights the way in which decision-making in experts is facilitated by an operator being able to draw on previous experience.

The use of simulation in decision-making training allows the careful construction of pre-rehearsed solutions to a range of problems encountered in complex real-world operations. In essence, this form of training seeks to develop a repository of basic experiences from which operators can quickly draw when solving common problems in the field.

In this type of training, it is not simply exposure to the scenario that is important in creating effective learning. Rather, the learning experience must also include careful facilitation in the processes underlying naturalistic decision-making, from guiding trainees to reflect on (1) the *plausible goals* of resolving the situation; (2) the *expectancies* in relation to how the situation will unfold; (3) the *critical cues* used in the on-going assessment and evaluation of the situation; and (4) the appropriate *courses of action* in response to the scenario.

Reinforcing the Use of Mnemonics

Another approach to the use of the simulator in decision-making training is to reinforce the use of mnemonics in the resolution of complex ill-defined problems. As described in the opening of this chapter, in some complex situations, operators cannot always draw on their expertise to quickly determine the appropriate course of action. In these situations, a problem cannot be resolved entirely by drawing on their repository of experience and pre-rehearsed solutions. In this case, operators must revert to the classical rational models of decision-making to diagnose the situation, and often derive possible solutions through inductive reasoning. In these situations, mnemonics such as A-GRADE, discussed earlier in this chapter, are used to structure and optimise the process of rational decision-making.

These forms of training, where the focus is on the *process* of rational decision-making rather than developing a repository of experiences, demands a completely different approach to scenario design. The focus of scenario design in this type of training must be on creating a complex ill-defined situation, for which there are multiple possibilities in terms of both the nature of the problem itself and the potential avenues the individual or team might take to resolve the situation.

One common misconception relating to these forms of decision-making training is that they must simply ensure that the individual or team is familiar with the mnemonic and follows each step in sequence. This approach fails to adequately support the more complex decision-making processes that are meant to be generated through the use of the mnemonic.

A more sophisticated approach involves using the mnemonic as a structure to promote complex inductive reasoning processes through which solutions can be generated for ill-defined and unexpected events. Recent research has highlighted the ways in which operators in certain high-risk industries, such as deep-sea fisheries, rely on decision-making as their primary risk management strategy rather than maintaining safety through the constraints of prescriptive rules and procedures.[21] Although somewhat beyond the scope of this book, training in making these types of decisions can be seen to support the individual and team processes that form part of engineering resilience into high-risk industries and promoting the innovation required to respond to unanticipated and almost unimaginable events.[22,23]

Scenarios Designed to Induce Negative Heuristics and Biases

Another important feature of scenario design for decision-making training using simulation is the ability to include in a scenario specific elements that may induce negative heuristics and biases.

In this form of training scenario, the simulator freeze technique can be used to great advantage to stop the unfolding events and prompt trainees to engage in critical metacognitive activities. As described earlier in this chapter, the only effective way to de-bias decision-making is to momentarily pause and reflect on the decision-making process to attempt to identify the influence of biases. Table 8.4 provides a summary of useful scenario design elements and metacognitive strategies to prompt trainees in developing the skills for de-biasing their decision-making processes.

Table 8.4 Heuristics and Biases that Influence Decision-Making

HEURISTIC/BIAS	SIMULATION TECHNIQUES
Availability heuristic	*Scenario design:* Creating a situation where the underlying problem is *not* the most common or likely event. *Metacognitive prompt:* Promoting trainees to consider a range of different, and unlikely, problems.
Confirmation bias	*Scenario design:* Only providing information to trainees that would confirm their initial hypothesis. *Metacognitive prompt:* Promoting trainees to identify sources of information that might indicate that their initial hypothesis might be incorrect.
Commission bias	*Scenario design:* In simulation-based training, trainees are naturally inclined to take action. Design a scenario where doing nothing is a legitimate immediate course of action. Also, a scenario where immediate action can exacerbate the situation is useful. *Metacognitive prompt:* Promoting trainees to reflect on what would have happened if they had not taken immediate action.
Search satisficing	*Scenario design:* This can easily be primed by a confederate agreeing with the first hypothesis and suggesting or commencing the logical first action given that hypothesis. *Metacognitive prompt:* As with confirmation bias, promoting trainees to identify sources of information that might indicate that their initial hypothesis might be incorrect.
Decision momentum	*Scenario design:* As above, a confederate prompting or undertaking continued actions can easily prime this bias. *Metacognitive prompt:* Prompting trainees to further consider alternative hypotheses or alternative courses of action.
Overconfidence bias	*Scenario design:* This bias can be primed by a confederate cued to subtly flatter and praise any erroneous aspects of the decision-making process, such as misdiagnosis or the suggestion of incorrect solutions. *Metacognitive prompt:* Prompting trainees to further consider alternative hypotheses or alternative courses of action and to reflect on whether they had made any errors.
Optimism bias	*Scenario design:* Creating a situation where information is consistent with a minor problem rather than the real and less pleasant problem. *Metacognitive prompt:* Prompting trainees to reconsider their initial diagnosis of the problem.

Creating Complex and High-Pressure Situations

In high-risk industries, critical decision-making often occurs in situations characterised by a challenging sub-optimal decision-making environment. For instance, a threat to safety often creates a situation of compressed time availability, high stress and high workload. Operators who must make decisions under these conditions need to have exposure to these conditions to build resilience and rehearse their non-technical skills while 'under fire'.[24] The simulator offers a unique environment in which these conditions can be created in a psychologically safe manner, and with the use of effective briefing and debriefing, effective training interventions can take place.

Developing and Maintaining Shared Mental Models

As we have explored, one of the most critical elements of effective decision-making in a team environment is the development and maintenance of a shared mental model.[25] The controlled environment of the simulator is perfect for rehearsing and coaching these aspects of team performance.

Decision-Making: On-the-Job Training

The coaching format of on-the-job training is an excellent mechanism for developing and enhancing skills relating to decision-making. Due to the intertwined aspects of technical knowledge and expertise in the decision-making process, training that deploys approaches found in the behavioural modelling, situated learning and cognitive apprenticeship models described in Chapter 3 is particularly useful for skill development in this domain of non-technical skills.

Modelling and Expertise

Workplace-based training for decision-making offers the opportunity for trainees to observe expert decision-making. However, one of the problems facing decision-making training relates to the fact that with expertise come more tacit and intuitive forms of decision-making, as has been previously discussed with respect to naturalistic decision-making. Therefore, for the novice, simple observation and modelling might not reveal much when an expert is operating in 'System 1' decision-making mode.

As one technique to address this issue, it is possible to repurpose some of the research tools used to understand expert decision-making. The most useful tool is the critical decision method (CDM), which was developed by Gary Klein as a tool to unmask the tacit knowledge used in expert decision-making.[26] The CDM asks an expert to reflect on a critical decision event and uses a series of probe questions to elicit tacit aspects of the decision-making process, such as cues, knowledge, analogous events, goals, options, the basis for choices and possible errors.[27]

When trainees are provided with a simplified version of the CDM, they have the tools to elicit an understanding of the decision-making processes used by experts in the workplace. This in turn provides a mechanism by which otherwise hidden aspects of expert decision-making can be observed and modelled.

Reinforcing Sources of Information and Resources

The types of decisions made in high-risk industries usually demand that multiple sources of information are sourced and integrated to understand the nature of the underlying problem and generate optimal solutions. Coaching efforts should therefore ensure that trainees are guided to all sources of information and resources for the decision-making process.

Decision-Making: Pre-Training Briefing

Briefing decision-making prior to training should focus on both the ways in which the decision-making process can be optimised and also the potential problems of decision-making facing both individuals and teams, such as heuristics and biases.

The pre-training briefing prior to a decision-making training session presents excellent opportunities for an integrated discussion of both the technical aspects of the problem to be encountered and also the non-technical skills involved in optimal decision-making processes.

AN EXAMPLE IN PRACTICE

The decision-making involved in the diagnosis and treatment of patients presenting at the emergency department is one of the most complex examples of decision-making in high-risk industries. A simulator scenario has been developed in which a trauma patient is admitted to the emergency department requiring resuscitation and treatment of complex internal injuries. The patient's management is made considerably harder by the addition of several significant comorbidities.

Prior to commencing the simulation, a briefing is undertaken that reviews the critical sources of information required for an optimal decision-making process, which integrates both technical and non-technical aspects. Then, a set of common decision-making biases are reviewed, and strategies that can be used by the team to avoid these biases are discussed.

Decision-Making: Assessment and Behavioural Markers

The assessment of decision-making skills needs to adopt a multi-faceted approach, which looks separately at the decision-making processes used by individuals and teams, the ways in which potential biases are mitigated and the eventual outcome of the decision-making process.

Avoiding Bias in Assessment Judgements

There can be considerable 'bleed' in assessment judgements when there is a mismatch between elements of performance. With decision-making, this is especially the case if the decision outcome is optimal but the process through which it was arrived at was sub-optimal. The assessor needs to be sure that their judgements of process and outcome are quite separate, and the debriefing can unpack the sub-optimal aspects of performance.

To achieve this, assessors should undertake de-biasing activities, as described in detail in Chapter 5 of this book.

Critical Decision Method

As was the case with situation awareness, it is quite easy to adapt methods used in researching decision-making processes to the task of assessing performance in the training environment. With respect to decision-making, an adaptation of the CDM provides a way to explore aspects of the decision-making process that might not be visible through observing performance alone. As described above, the CDM asks an expert to reflect on a critical decision event and uses a series of probe questions to elicit the tacit aspects of the decision-making process, such as cues, knowledge, analogous events, goals, options, the basis for choices, and possible errors.[27] The information made available through this process can then be used in forming judgements about the appropriateness and efficacy of the decision-making process during training and assessment.

Behavioural Markers

One of the most popular frameworks for assessing the non-technical skills of flight crew, NOTECHS, specifies four elements under the category of decision-making, as described in Table 8.5. Each of these four elements represents a sequential step in the overall decision-making process, which assists in structuring the assessment process and being able to focus in on different elements of decision-making in assessment and debrief.

As discussed in Chapter 5, it is also possible to create a more detailed set of behavioural markers that are specific to the training scenario being used. This is particularly effective in the decision-making domain, as each decision will be structured around scenario-specific cues, available options for courses of action, and associated risks. Table 8.6 provides an example set of behavioural markers, similar to those used to evaluate the decision-making performance of commercial airline pilots.[28]

Table 8.5 Typical Elements within the Domain of Decision-Making

CATEGORY	ELEMENT
	Problem definition/diagnosis
Decision-making	Option generation
	Risk assessment/option choice
	Outcome review

Table 8.6 Specific Behavioural Markers for a Weather-Related Decision

DECISION ELEMENT	DESCRIPTION
Landing distance	Did the crew correctly calculate the landing distance required given the weather conditions and aircraft configuration?
Alternate airport	Did the crew correctly identify the need for an alternate airport and choose a suitable airport in the case of diversion?
Weather updates	Did the crew obtain any updates on weather conditions at the destination?
Contingency plans	Did the crew identify and evaluate a range of options that would be available as contingency plans if conditions changed?

Decision-Making: Debriefing and Coaching

As with situation awareness, the debriefing of decision-making should focus on both the outcome of the decision and the processes used during decision-making.

In Chapter 6, we explored in detail the process of effective facilitated debriefing and emphasised the role of the instructor as an active listener. However, one of the challenges facing the instructor when facilitating the debrief of a decision-making exercise is that, according to our understanding of naturalistic decision-making, aspects of the decision-making process may well be tacit. That is to say, trainees may not be fully aware of how they went about the decision-making process. In this situation, the facilitator might need to use a much more probing and questioning approach to the debrief than might be the case with other non-technical skills. Trigger questions for debriefing decision-making might include the following:

- What were the most important pieces of information that triggered the need for a decision to be made?
- At this point in time, what did you think was the nature of the problem?
- What options did you think you had available?
- Did this situation remind you of anything else you have experienced in the past, and did this help you? (analogical reasoning)
- Were you aware of any biases that might have influenced your decision?
- With the benefit of hindsight, did you leave anything out of the decision-making process?

Decision-Making: Key Resources

The following are excellent resources that provide assistance in the process of designing effective decision-making training and assessment programs.

Gladwell, M. (2005). *Blink: The Power of Thinking without Thinking*. Boston, MA: Back Bay Books.

Kahneman, D. (2011). *Thinking, Fast and Slow*. London, UK: Penguin Books.

Klein, G. A. (1993). A recognition-primed decision (RPD) model of rapid decision making. In G. A. Klein, J. Orasanu, R. Calderwood and C. E. Zsambok (Eds.), *Decision Making in Action: Models and Methods* (pp. 138–147). Norwood, NJ: Ablex.

Klein, G. A. (1997). An overview of naturalistic decision making applications. In C. E. Zsambok and G. A. Klein (Eds.), *Naturalistic Decision Making* (pp. 49–60). Mahwah, NJ: Lawrence Erlbaum Associates.

Case Studies

US Airways Flight 1549: Often referred to as the 'miracle' on the Hudson, the successful ditching of the US Airways A320 after loss of thrust in both engines due to damage caused by hitting a flock of geese is a great example of decision-making under pressure. The National Transportation Safety Board (NTSB) concluded in its accident investigation that the captain's decision to ditch on the Hudson River rather than attempting to land at an airport provided the highest probability that the accident would be survivable.[29]

Piper Alpha Disaster: On the evening of 6 July 1988, there was an explosion on board the offshore oil platform Piper Alpha. A total of 167 lives were lost. The majority of these were not associated with the initial explosion but, rather, resulted from decision-making failures, both by the offshore installation manager with respect to crisis management and also due to the response from the linked Claymore platform to continue production.[30]

References

1. Wickens, C. D., Gordon, S. E., and Liu, Y. (1998). *An Introduction to Human Factors Engineering*. New York, NY: Addison Wesley Longman.
2. Gigerenzer, G., and Selten, R. (2002). *Bounded Rationality: The Adaptive Toolbox*. Boston, MA: MIT Press.

3. Simon, H. A. (1982). *Models of Bounded Rationality: Empirically Grounded economic reason* (Vol. 3). Boston, MA: MIT Press.
4. Kahneman, D. (2003). A perspective on judgment and choice: Mapping bounded rationality. *American Psychologist*, *58*(9), 697.
5. Klein, G. A. (1997). An overview of naturalistic decision making applications. In C. E. Zsambok and G. A. Klein (Eds.), *Naturalistic Decision Making* (pp. 49–60). Mahwah, NJ: Lawrence Erlbaum Associates.
6. Harrison, Y., and Horne, J. A. (2000). The impact of sleep deprivation on decision-making: A review. *Journal of Experimental Psychology: Applied*, *6*, 236–249.
7. Klein, G. A. (1993). A recognition-primed decision (RPD) model of rapid decision making. In G. A. Klein, J. Orasanu, R. Calderwood and C. E. Zsambok (Eds.), *Decision Making in Action: Models and Methods* (pp. 138–147). Norwood, NJ: Ablex.
8. Fogarty, W. M. (1988). *Formal Investigation into the Circumstances Surrounding the Downing of Iran Air Flight 655 on 3 July 1988*. Arlington, VA: Department of Defense.
9. Gladwell, M. (2005). *Blink: The Power of Thinking without Thinking*. Boston, MA: Back Bay Books.
10. Kahneman, D. (2011). *Thinking, Fast and Slow*. London, UK: Penguin Books.
11. Croskerry, P. (2002). Achieving quality in clinical decision making: Cognitive strategies and detection of bias. *Academic Emergency Medicine*, *9*(11), 1184–1204.
12. Salas, E., Sims, D. E., and Burke, C. S. (2005). Is there a 'big five' in teamwork? *Small Group Research*, *36*(5), 555–599.
13. Duffy, L. (1993). Team decision making biases: An information processing perspective. In G. A. Klein, J. Orasanu, R. Calderwood and C. E. Zsambok (Eds.), *Decision Making in Action: Models and Methods* (pp. 346–359). Norwood, NJ: Ablex.
14. Bearman, C., Paletz, S. B. F., Orasanu, J., and Thomas, M. J. W. (2010). The breakdown of coordinated decision making in distributed systems. *Human Factors*, *52*(2), 173–188.
15. Cannon-Bowers, J. A., and Bell, H. H. (1997). Training decision makers for complex environments: Implications of the naturalistic decision making perspective. In C. E. Zsambok and G. A. Klein (Eds.), *Naturalistic Decision Making* (pp. 99–110). New York, NY: Psychology Press.
16. Edkins, G. D. (2000). Mind over Matter. *Flight Safety Australia, September*, 31–33.
17. Croskerry, P. (2003). Cognitive forcing strategies in clinical decision making. *Annals of Emergency Medicine*, *41*(1), 110–120.
18. Croskerry, P. (2003). The importance of cognitive errors in diagnosis and strategies to minimize them. *Academic Medicine*, *78*, 1–6.
19. Tawfik, A., and Jonassen, D. (2013). The effects of successful versus failure-based cases on argumentation while solving decision-making problems. *Educational Technology Research and Development*, *61*(3), 385–406.

20. Taleb, N. N. (2007). *The Black Swan: The Impact of the Highly Improbable*. New York, NY: Random House.
21. Morel, G., Amalberti, R., and Chauvin, C. (2008). Articulating the differences between safety and resilience: The decision-making process of professional sea-fishing skippers. *Human Factors*, *50*(1), 1–16.
22. Hollnagel, E., Nemeth, C. P., and Dekker, S. W. A. (2008). *Resilience Engineering Perspectives: Remaining Sensitive to the Possibility of Failure*. Aldershot, UK: Ashgate Publishing.
23. Hollnagel, E., Woods, D. D., and Levenson, N. (2006). *Resilience Engineering: Concepts and Precepts*. Aldershot, UK: Ashgate Publishing.
24. Cannon-Bowers, J. A., and Salas, E. (1998). *Making Decisions under Stress: Implications for Individual and Team Training*. Washington, DC: American Psychological Association.
25. Stout, R. J., Cannon-Bowers, J. A., Salas, E., and Milanovich, D. M. (1999). Planning, shared mental models, and coordinated performance: An empirical link is established. *Human Factors*, *41*(1), 61–71.
26. Klein, G. A., Calderwood, R., and MacGregor, D. (1989). Critical decision method for eliciting knowledge. *IEEE Transactions on Systems, Man and Cybernetics*, *19*(3), 462–472.
27. Hoffman, R. R., Crandall, B., and Shadbolt, N. (1998). Use of the critical decision method to elicit expert knowledge: A case study in the methodology of cognitive task analysis. *Human Factors: The Journal of the Human Factors and Ergonomics Society*, *40*(2), 254–276.
28. Petrilli, R. M., Thomas, M. J. W., Lamond, N., Dawson, D., and Roach, G. D. (2007). Effects of flight duty and sleep on the decision-making of commercial airline pilots. In J. M. Anca (Ed.), *Multimodal Safety Management and Human Factors* (pp. 259–270). Aldershot, UK: Ashgate Publishing.
29. Glazner, L. K. (1991). Shiftwork: Its effect on workers. *AAOHN Journal*, *39*(9), 416–421.
30. Flin, R. (2001). Decision making in crises: The Piper Alpha disaster. In U. Rosenthal, A. Boin and L. K. Comfort (Eds.), *Managing Crises: Threats, Dilemmas, Opportunities* (pp. 103–118). Springfield, IL: Charles C Thomas Publisher.

9

Training and Assessing Communication and Teamwork Skills

Communication and Teamwork: A Primer

Effective communication and teamwork are essential for safe operations across all high-risk industries. As humans are an inherently social species, and a species with complex verbal and non-verbal communication abilities that develop from infancy through to adulthood, the non-technical skills domain of communication and teamwork has sometimes suffered from being largely ignored in formal training interventions.[1] Instead, the skills in this domain are often seen as being brought to the work environment and are often the focus of modern assessment techniques in employee selection across many high-risk industries.

High-risk work environments have very specific demands with respect to communication and teamwork. Given the nature of operations in these industries, they can be truly unforgiving of failures in communication and of ineffective teamwork. In recent years, the continued influence of poor communication and teamwork in catastrophic events in high-risk industries has highlighted that skills in this domain do need to be the focus of formal training programs, and the lack of attention given to communication skills in *ab initio* training needs to be addressed and rectified.

Communication and Teamwork: Exemplar Case Study

On 6 July 1988, 167 people perished as a result of a massive fire on the Piper Alpha offshore platform in the North Sea oil and gas field. The context of this accident involved a large body of planned maintenance works on the platform, which were to be undertaken while the platform was still operating. Part of these maintenance activities involved regular on-going servicing of critical parts. The platform had some 300 pressure safety valves, each of which had to be recertified by a specialist contractor every 18 months.[2] Other, more major works were occurring at the time, and one of the two main condensate pumps on the platform, Pump A, had been prepared to be removed from service as part of these activities.

During the day shift on 6 July, a contractor valve technician had undertaken works on one of Pump A's pressure safety valves. Not having been able to complete the work by the end of the day shift, they fitted a blank flange where the pump had been removed, in preparation for refitting the valve the following day. This blank flange was not intended to be leak tight, as it was expected that the pump would not be operated.

However, not long into the night shift, an issue with the other main condensate pump, 'Pump B', became evident. To maintain operation of the Piper Alpha facility, a solution to this problem had to be found. The control room personnel were aware that Pump A had been prepared for major works, but they were able to determine from the 'Permit to Work' system that those works had not yet commenced. However, they were unaware that a contractor had been working on the pressure safety valve of Pump A and that the pump was unserviceable.

Without this information, the night shift operators decided that they could start Pump A, unaware that the pressure safety valve had been removed and replaced with a blank flange, which was not pressure tight. The lead night shift operator started Pump A, immediately causing a leak of condensate at the point of the blank flange. This condensate subsequently ignited, and the resulting explosion triggered the catastrophic series of fires that effectively destroyed the platform.[3] Compounding the issue was the fact that Piper Alpha was not isolated from two nearby platforms for over an hour, with those platforms effectively feeding oil to fuel the fire on Piper Alpha.

This accident is an exemplar for multiple communication failures leading to a catastrophic outcome and highlights a number of areas where communication can be a significant point of vulnerability in high-risk industries.

The first main area where breakdowns in communication occurred was at the interfaces between work-groups and between shifts. In high-risk industries, we use the term *handover* to describe the communication that occurs at these interfaces.[4] There is considerable evidence to support the fact that these handovers are points of vulnerability.[5] In the Piper Alpha event, there were three specific handovers between lead process operators, the condensate area operators and the maintenance lead operators, where it should have been communicated that the relief valve had been removed and had not yet been replaced.[3]

The second main area where breakdowns in communication occurred related to the written communication systems where records of work activities were recorded. In the oil and gas industry, as is the case in all production and extractive industries, a 'Permit to Work' system was in place to provide written records of the maintenance status of parts of the facility and what pieces of equipment were 'locked out' for maintenance purposes. These forms of written communication are commonplace across high-risk industries, particularly in healthcare, where the patient's notes, medication chart and other written forms of communication are absolutely critical in the provision of care. Each of these systems enables the communication of safety-critical information in a manner that does not rely on human memory and can be used across and between teams to ensure continuity of operations.

Communication and Teamwork: Core Enabling Knowledge

Even though communication is one of our very basic foundation human skills, much of what we have learnt by way of communication techniques are forms of tacit knowledge. Therefore, it is necessary to build an explicit understanding of human communication if we are to develop enhanced communication skills.

Communication can be defined as the process of exchanging information, ideas and feelings.[6] Communication can take place between two people or within a group. In high-risk industries, communication is closely aligned with the notion of teamwork, and we shall explore these two aspects of non-technical performance together.

Models of Communication

An obvious place to begin the development of core enabling knowledge relating to communication and teamwork is through exploring the basic models of communication. The most common of these involves a process of a person encoding a message into words or other forms and transmitting that message to one or more people, who receive, decode and make sense of the message. This model can then be translated from one-to-one communication to one-to-many or many-to-many forms of communication within a team environment.

Modes of Communication: Verbal, Intonation, Non-Verbal and Written

Communication can use a variety of modes, and to make matters even more complicated, these modes of communicating can be used separately or on their own. Our most common form of communication is verbal in nature, and the fact that we have developed complex language skills is one of the features that set us apart as human.

As a species, we have developed sophisticated forms of non-verbal communication, which can augment our verbal communication or be used as stand-alone forms of communication. As many of our high-risk work environments are distributed, and with safety-critical communication being exchanged via text or transmitted via radio or telephone, the absence of non-verbal cues needs to be acknowledged and compensatory strategies employed. For instance, we frequently use non-verbal cues for emphasis or to convey a sense of importance or urgency. Communication stripped of these cues must therefore use other techniques to convey these contextual elements.

Information Exchange and Closed Loop Communication

Skills in the effective packaging and transmission of information and techniques involved in closed loop coordination form perhaps the most basic principles of communication in high-risk industries. The critical role of standard phraseology and techniques for simplifying and enhancing the clarity of communication is essential in high-risk industries.

Similarly, techniques for making the communication clear through emphasis, intonation, and the role of keywords and non-verbal cues in expressing a sense of urgency are all important elements of knowledge that can very easily be put into practice through more active modes of learning.

The critical role of read-back and other techniques of closed loop communication as mechanisms to ensure that the message was received and understood as intended also forms a basic aspect of knowledge about communication in high-risk industries.

Barriers to Effective Communication and Teamwork

Within the context of high-risk industries, a range of factors can be seen as presenting barriers to effective communication and teamwork. Many of these relate to the fact that operations in high-risk industries are frequently undertaken by teams made up of individuals from different disciplines; for instance, the surgeon, anaesthetist and nurses as the core team in the operating room.

To achieve optimal performance in high-risk industries, each of these barriers must be overcome. To this end, specific and often formalised communication protocols, techniques and strategies have been developed to enhance communication (Table 9.1).

Enquiry and Assertiveness

One of the most critical communication skills in a high-risk environment is that of enquiry. It may sound obvious, but the ability to seek

Table 9.1 A Summary of Barriers to Effective Communication in High-Risk Industries

BARRIER	DESCRIPTION
Cultural	Often, teams are comprised of individuals from different disciplines, which bring their own historical professional cultures. The professional cultures, when misaligned, can present barriers to effective communication and teamwork.[7]
Technical	Often, the different disciplines have very different technical knowledge and expertise. This makes communication about non-normal situations more difficult due to constraints in establishing a shared mental model and in using appropriate terminology.
Physical	The physical barriers that are present in any distributed teamwork can constrain communication and present barriers to effective teamwork.[8]
Organisational	Having teams comprised of individuals from separate departments in an organisation can also present barriers to effective communication and teamwork.[8]
Hierarchical	Institutionalised hierarchies, such as chain of command, can present significant barriers to communication, as junior team members often feel unable to speak up when something is potentially wrong.

clarification in the instance of any doubt is critical to the processes of maintaining a shared mental model and also error detection. Training can emphasise instances where someone has been concerned but has not spoken up, and as a consequence, safety has been compromised.

Team Coordination

Effective communication is also critical in the process of team coordination to ensure safe and efficient operations in high-risk industries.

Communication and Teamwork: Skill Development

As with other non-technical domains, the development of communication skills must be focussed on both the individual and the team. Early in the history of crew resource management in aviation, it was identified that a set of important communication skills such as enquiry, assertiveness and conflict resolution could be enhanced through dedicated training.[9]

Exchanging Information

The history of catastrophic events in high-risk industries is riddled with examples of situations where critical information, often known by one or more team members, is not communicated in a timely fashion.

As with many aspects of non-technical skills, mnemonics are frequently used as memory prompts for effective communication strategies. In healthcare, one of the most ubiquitous of these is ISOBAR and its derivatives. This mnemonic has been designed to be used in several different situations, from junior staff members seeking advice from senior colleagues, through to the handover of patient care from one team to another. Table 9.2 provides an overview of the ISOBAR mnemonic.[10]

Enquiry and Graded Assertiveness

Another critical communication skill that can be developed in non-technical skills training programs is that of enquiry and assertiveness.

Table 9.2 The ISOBAR Mnemonic

ISOBAR	DESCRIPTION
Identify	Introduce yourself and the patient to ensure positive patient identification.
Situation	Describe the current situation to provide some brief context.
Observations	Convey all relevant vital signs, test results and a clinical assessment of the patient.
Background	Put this information in the context of the patient's relevant clinical history.
Agreed plans	Determine what needs to happen.
Read-back	The plan is read back to ensure that joint understanding has been achieved and nothing has been omitted from the plan.

Perhaps one of the most important non-technical skills to be developed in high-risk industries is the ability to speak up and show concern at appropriate times. Assertiveness training has been used as a countermeasure to cultures (national and organisational as well as professional cultures) where a steep authority gradient might present a barrier to junior team members speaking up.[7,11]

Graded assertiveness is a technique whereby different levels of assertiveness (from low to high) are used to escalate a situation that is not being effectively managed. One mnemonic used in this practice is 'probe, alert, challenge, emergency' (PACE). When someone is unsure about a situation unfolding, they begin with a gentle *probe* that is often phrased as a question, such as 'what was our plan in relation to …?', or a hint, 'did you know that …?' If this does not prompt the necessary response, the next stage is an alert, often framed as raising concern about the situation. Then, if this does not prompt the necessary response, the next stage is to issue a direct challenge to the situation, highlighting explicitly what the alternative actions or solutions should be. Finally, if none of these previous stages has prompted action, the final stage is to declare an *emergency* situation in which all current activities must stop.[12] Table 9.3 provides an overview of the PACE mnemonic.

Table 9.3 The PACE Mnemonic

PACE	DESCRIPTION
Probe	A hint or gentle suggestion that something might be wrong.
Alert	Highlighting your concern for what is occurring.
Challenge	Making a specific suggestion to stop what is being done, or to suggest an alternative.
Emergency	Using emergency language such as 'Stop – you must listen.'

Even going beyond training in graded assertiveness techniques, some organisations have formalised the process through specific standard operating procedures, where each stage of graded assertiveness is mapped out, and the final stage is specified as a very specific phrase such as 'Captain, you *must* listen!' If the emergency language is used, there is a formal requirement for subsequent action and reporting.

Team Coordination Training

There is a set of specific skills associated with coordination among team members. In many high-risk industries, the traditional model of a single person being in charge is largely irrelevant. Examples of this include the coordination between flight crew and air traffic control, the coordination between a control room operator and a worker in the field, and the coordination among a healthcare team during a trauma resuscitation. When coordination in these environments breaks down, safety can easily be compromised.[13,14]

Team coordination involves authentic scenario-based training in which teams are required to practice the delegation and coordination of tasks to ensure that performance is optimised for safety and efficiency. As is now the case with many aspects of non-technical skills, there is now a body of scientific evidence to suggest that participation in such training leads to enhanced team behaviours, a reduction in consequential errors, and improved staff attitudes towards teamwork concepts.[15]

Team Self-Correction Training

High-performing teams are ones in which error is anticipated and members of the team are able to identify and correct errors. In this type of environment, errors are owned by the whole team and not seen simply as deficient performance of one team member. Rather than the culture of the team making individuals hesitant to highlight another team member's error, there is actually an obligation to do so. A perfect example of this working in practice is on the flight deck of a commercial airliner. Over the years since the introduction of crew resource management (CRM) in aviation, the best-performing flight decks display an expectation that error will occur, and a pilot will almost feel let down if they make an error and they are not alerted to it by the person sitting next to them.

Training team self-correction is easily achieved by the instructor modelling appropriate actions in the training process and then gradually removing this support and allowing teams to self-correct.[16] Again, team correction training can be supported through formalised processes of cross-checking or cross-monitoring and through the use of simple mnemonics such as 'I monitor, I check, we correct.'[17]

Communication and Teamwork: Simulation-Based Training

The authentic environment of the simulator, where teams can work together on anything from an everyday procedure through to an emergency situation, lends itself entirely to the development of communication and teamwork skills.

Low Fidelity and Role-Play

Unlike the domains of decision-making and situation awareness, where there is a very clear need for training to be contextualised in authentic real-world problem spaces, the development of communication skills is one area that can harness lower-fidelity forms: simulation and role-play.

One classic example of the use of role-play for training communication skills is a scenario developed for sensitive communication around the issue of managing a colleague impaired by drugs or alcohol. Instead of using the costly resources of high-fidelity simulation, an instructional decision was made to tackle this issue as a role-play exercise introduced unexpectedly in the pre-briefing stage of an anaesthetic crisis resource management training session. With a confederate playing the role of an impaired colleague, the other participants in the training session needed to effectively and sensitively manage the situation. Using the trial and error approach introduced in Chapter 3, participants were given the opportunity to manage the situation; then, the 'scenario' was stopped, the communication strategies were debriefed, and participants then repeated the scenario, avoiding previous errors and rehearsing effective strategies for sensitive communication with an impaired colleague. To this day it remains a very powerful moment of learning for all involved.

Confederates: Scripting Roles to Elicit Specific Strategies

The simulator presents unique opportunities with respect to designing scenarios that require specific communication and teamwork strategies to be developed and practised. Often, this is achieved by including scripted roles in the scenario design, such as the surgeon who won't listen or the captain who continues in an unsafe situation such as an unstable approach. These scripted roles are often referred to as *confederates* in the scenario design.[18]

Challenging Cultural Norms and Legitimising New Norms

Often, some aspects of communication skills, such as assertiveness, are difficult to develop due to the way in which they might challenge cultural norms. For instance, a junior doctor or nurse may have difficulty speaking up about a potentially dangerous decision by a senior doctor due to the effects of hierarchy.

The simulator provides a psychologically safe environment in which to practise behaviour that might be perceived to challenge these cultural norms. Moreover, through the process of briefing, practice and debriefing, the otherwise uncomfortable communication behaviour becomes legitimised.

Communication and Teamwork: On-the-Job Training

The development of communication skills and teamwork strategies on the job presents a number of challenges. As communication is a ubiquitous part of human endeavour, it is difficult to tailor situations where actual work tasks can be performed and the communication and teamwork elements can also be constantly reflected on. However, a number of techniques can be drawn from how we approach training non-technical skills in the simulator and applied to on-the-job training of communication and teamwork skills.

The 'Freeze' Technique

Although it is usually reserved for use in simulation-based training, a supervisor or coach can also use the 'freeze' technique during on-the-job training. This involves a task being momentarily stopped and aspects of communication and teamwork subjected to facilitated debrief and critical reflection. Changes can then be made to communication processes or other elements of how the team is working together, and then the work task can be recommenced. However, the supervisor or coach needs to be sensitive to the fact that this technique is in effect an interruption, and steps need to be taken to ensure that interrupting a job part way through does not compromise safety.

Video-Based Debrief

As we are often not entirely aware of our patterns of communication and our non-verbal forms of communication, the use of video is an important instructional tool for training communication and teamwork skills. Although not suitable in all work environments, the use of video to assist in post-work facilitated debriefing can result in great learning moments.

Critical Incident Debrief

A final technique that can be used for on-the-job training of communication and teamwork skills is the traditional critical incident debrief. After a significant event has occurred, a supervisor or coach can help debrief the incident and facilitate reflection on how communication and teamwork might have contributed to good and sub-optimal aspects of how the critical incident was managed.

Communication and Teamwork: Pre-Training Briefing

As we have seen throughout this book, the briefing undertaken prior to training should focus on three major aspects: the expectations and desired outcomes of the training event, a review of the underlying enabling knowledge, and then a more detailed discussion of the specific skills that are to be used and rehearsed during in the training event.

As communication is a particularly broad set of non-technical skills, the briefing should focus on isolating the key communication and teamwork skills relevant to the training scenario. For instance, if the scenario has been designed to elicit assertiveness, then the underlying principles, general techniques, and any specific strategies or mnemonics used in the organisation to support assertiveness should all be highlighted in the pre-training briefing. All these features of the briefing facilitate the activation of core enabling knowledge, reinforcing the key skills and techniques and priming trainees for learning.

If the training event has been designed around a complex scenario in high-fidelity simulation, the briefing can provide an opportunity for a brief and simple role-play exercise to be used as a method for initial practice prior to demonstrating the application of the relevant skills in the simulator. Many times, learning has been inhibited by a trainee being thrown into the simulator in the 'hot seat' and their communication or team coordination performance being poor directly due to inadequate preparatory briefing of the skills and strategies that are needed for optimal performance.

AN EXAMPLE IN PRACTICE

In aviation, the classic pre-training briefing of a complex flight manoeuvre such as circling to land at an airport after a non-precision approach often involves the flight path being drawn on a whiteboard, with the instructor facilitating crew discussion of critical points in the approach sequence with respect to aircraft altitude, speed and configuration.

Training a critical incident management team in the skill of team coordination can be done in very much the same way. A timeline of events can be mapped out on the whiteboard, and the briefing can involve facilitated discussion of the various roles and responsibilities in the team, who should be doing what at various points in time, who needs to be conveying critical information, and how the team can be coordinated.

Communication and Teamwork: Assessment and Behavioural Markers

As the majority of performance relating to communication and teamwork is directly observable, it presents significantly fewer challenges to assessment than the non-technical skills of situation awareness and decision-making. Given that communication and teamwork are observable in this way, the use of behavioural markers is the most important tool for assessing this domain of non-technical skill.

The domain of communication and teamwork is quite broad, and as such, a range of different elements, or constituent skills, have been identified. Table 9.4 highlights a range of these constituent skills.[19,20]

Specific behavioural markers can then be developed for each of the elements as they relate to the specific job-roles, tasks or scenarios. As discussed in Chapter 5, consideration needs to be given here to the degree of generality or specificity described in the set of behavioural markers. While effective communication can be described in broad and general terms, elements of teamwork are highly context specific and may lend themselves to being described in a much more scenario-dependent manner. Table 9.5 provides an example of a set of behavioural markers for the skill of using authority and assertiveness as described in the anaesthetists' non-technical skills (ANTS) framework.[19]

Table 9.4 Typical Elements within the Domain of Communication and Teamwork

CATEGORY	ELEMENT
	Communication environment
Communication and teamwork	Sharing information
	Assertiveness
	Enquiry
	Coordination of activities
	Supporting others

Table 9.5 Example Behavioural Markers for Communication and Teamwork Element 'Using Authority and Assertiveness' in the ANTS Framework

ELEMENT	BEHAVIOURAL MARKERS FOR GOOD PRACTICE
Using authority and assertiveness	• Makes requirements known with necessary level of assertiveness • Takes over task leadership as required • Gives clear orders to team members • States case and provides justification
	BEHAVIOURAL MARKERS FOR POOR PRACTICE
	• Does not challenge senior colleagues or consultants • Does not allow others to put forward their case • Fails to attempt to resolve conflicts • Does not advocate position when required

Source: Fletcher, G. et al., *British Journal of Anaesthesia*, *90*(5), 580–588, 2003.

Communication and Teamwork: Debriefing and Coaching

Debriefing and coaching communication and teamwork need to focus not only on the specific behaviours of individuals but also on how other team members responded to those behaviours. In other words, the perspectives of both the sender and the receiver need to be explored.

Video is an extremely powerful learning tool in the debriefing of training relating to communication and teamwork. The look of shock on participants' faces when they hear back their actual communication during a high-workload emergency situation is often enlightening. Powerful debriefs with video often highlight communication errors, such as when someone asks for something critical to be obtained, but the request evaporates into thin air because no one has been named as being assigned to that task. Similarly, instances of unclear or ambiguous information exchange can be highlighted extremely well through video replay. Some key trigger questions for debriefing communication and teamwork might include the following:

- How well did you feel you were working as a team?
- How did you feel you communicated that critical piece of information?
- When you raised that concern, did you feel you were listened to?
- Do you think you were given enough information to fully appreciate the situation?
- Did you feel you understood your role and what others in the team had to achieve?
- Were there any points in time when you felt you needed clearer guidance?

Communication and Teamwork: Key Resources

The following books provide an excellent overview of elements of communication and teamwork across high-risk industries:

Flin, R. H., O'Connor, P., and Crichton, M. (2008). *Safety at the Sharp End: A Guide to Non-Technical Skills*. Aldershot, UK. Ashgate Publishing.

Salas, E., Bowers, C. A., and Edens, E. (Eds.). (2001). *Improving Teamwork in Organizations: Applications of Resource Management Training*. Mahwah, NJ: Lawrence Erlbaum Associates.

Case Studies

Continental Express Flight 2574: In September 1991, a Continental Express Embraer broke up in flight en route to Houston, Texas. The investigation revealed that during maintenance, the screws that secured the top of the leading edge of the horizontal stabiliser had been removed but not replaced. This occurred because a maintenance inspector had attempted to assist two mechanics in a job that required removal of the leading edge of the horizontal stabiliser. The inspector worked on the top, while the two mechanics worked on the bottom. On shift handover, it was communicated that only the right-hand side had been worked on, whereas in actuality the inspector had removed both left- and right-side screws on the top of the stabiliser. This was not visible from inspection on the ground. Work was completed on the right-hand side by the next shift, but the screws were never replaced on the left-hand side, resulting in the leading edge tearing from the aircraft in flight, which led to an in-flight breakup.[21]

Air Ontario Flight 1363: On 10 March 1989, an Air Ontario F-28 was preparing to depart Dryden on a snowy day. This accident involved cabin crew not conveying to the flight crew the concerns of a travelling pilot about the degree of snow and ice accumulation on the aircraft's wings. This snow and ice subsequently led to the aircraft not being able to attain sufficient altitude at the end of the runway. After this accident, many airlines developed joint CRM training programs, where both flight crew and cabin crew come together for training.

References

1. Leonard, M., Graham, S., and Bonacum, D. (2004). The human factor: The critical importance of effective teamwork and communication in providing safe care. *Quality and Safety in Health Care*, *13*(suppl 1), i85–i90.
2. Reason, J., and Hobbs, A. (2003). *Managing Maintenance Error: A Practical Guide*. Aldershot, UK: Ashgate Publishing.
3. Appleton, B. (2001). Piper Alpha. In T. Kletz (Ed.), *Learning from Accidents* (pp. 196–206). Oxford, UK: Gulf Professional Publishing.
4. Catchpole, K. R., De Leval, M. R., Mcewan, A., Pigott, N., Elliott, M. J., Mcquillan, A., Macdonald, C., and Goldman, A. J. (2007). Patient handover from surgery to intensive care: Using Formula 1 pit-stop and aviation models to improve safety and quality. *Pediatric Anesthesia*, *17*(5), 470–478.
5. Thomas, M. J. W., Schultz, T. J., Hannaford, N., and Runciman, W. (2013). Failures in transition: Learning from incidents relating to clinical handover in acute care. *Journal of Healthcare Quality*, *35*(3), 49–56.
6. Flin, R. H., O'Connor, P., and Crichton, M. (2008). *Safety at the Sharp End: A Guide to Non-Technical Skills*. Aldershot, UK: Ashgate Publishing.

7. Helmreich, R. L., and Merritt, A. C. (1998). *Culture at Work in Aviation and Medicine: National, Organizational and Professional Influences*. Aldershot, UK: Ashgate Publishing.
8. Chute, R. D., and Weiner, E. L. (1995). Cockpit-cabin communication: I. A tale of two cultures. *The International Journal of Aviation Psychology, 5*(3), 257–276.
9. Jensen, R. S., and Biegelski, C. S. (1989). Cockpit resource management. In R. S. Jensen (Ed.), *Aviation Psychology*. Aldershot, UK: Gower.
10. Porteous, J. M., Stewart-Wynne, E. G., Connolly, M., and Crommelin, P. F. (2009). ISOBAR: A concept and handover checklist – The National Clinical Handover Initiative. *Medical Journal of Australia, 190*(11), S152–S156.
11. Helmreich, R. L., Wilhelm, J. A., Klinect, J. R., and Merrit, A. C. (2001). Culture, error and crew resource management. In E. Salas, C. A. Bowers and E. Edens (Eds.), Improving Teamwork in Organizations (pp. 305–331). Mahwah, NJ: Lawrence Erlbaum Associates.
12. Lancman, B., and Jorm, C. (2015). Taking the heat in critical situations: Being aware, assertive and heard. In R. Iedema, D. Piper and M. Manidis (Eds.), *Communicating Quality and Safety in Health Care* (pp. 268–278). Cambridge, UK: Cambridge University Press.
13. Owen, C., Bearman, C., Brooks, B., Chapman, J., Paton, D., and Hossain, L. (2013). Developing a research framework for complex multi-team coordination in emergency management. *International Journal of Emergency Management, 9*(1), 1–17.
14. Bearman, C., Paletz, S. B. F., Orasanu, J., and Thomas, M. J. W. (2010). The breakdown of coordinated decision making in distributed systems. *Human Factors, 52*(2), 173–188.
15. Morey, J. C., Simon, R., Jay, G. D., Wears, R. L., Salisbury, M., Dukes, K. A., and Berns, S. D. (2002). Error reduction and performance improvement in the emergency department through formal teamwork training: Evaluation results of the MedTeams project. *Health Services Research, 37*(6), 1553–1581.
16. Wilson, K. A., Salas, E., Priest, H. A., and Andrews, D. (2007). Errors in the heat of battle: Taking a closer look at shared cognition breakdowns through teamwork. *Human Factors,* 49(2), 243–256.
17. Risser, D. T., Simon, R., Rice, M. M., and Salisbury, M. L. (1999). A structured teamwork system to reduce clinical errors. In P. L. Spath (Ed.), *Error Reduction in Health Care: A Systems Approach to Improving Patient Safety* (pp. 230–240). San Francisco, CA: Jossey-Bass.
18. Orasanu, J., Fischer, U., McDonnell, L. K., Davison, J., Haars, K. E., Villeda, E., and VanAken, C. (1998). How do flight crews detect and prevent errors? Findings from a flight simulation study. Paper presented at the Proceedings of the Human Factors and Ergonomics Society annual meeting.

19. Fletcher, G., Flin, R., McGeorge, P., Glavin, R. J., Maran, N. J., and Patey, R. (2003). Anaesthetists' non-technical skills (ANTS): Evaluation of a behavioural marker system. *British Journal of Anaesthesia*, *90*(5), 580–588.
20. Klampfer, B., Flin, R., Helmreich, R. L., Häusler, R., Sexton, B., Fletcher, G., Field, P., et al. (2001). *Enhancing Performance in High Risk Environments: Recommendations for the Use of Behavioural Markers*. Ladenburg: Daimler-Benz Stiftung.
21. NTSB. (1992). *Continental Express Flight 2574 In-Flight Structural Breakup Emb120RT, N33701 Eagle Lake, Texas September 11,1991*. Washington, DC: National Transport Safety Board.

10

Training and Assessing Task Management

Task Management: A Primer

The domain of task management is fundamental to the effective organisation of work in the complex and dynamic environments of high-risk industries. History tells us that it is not only during crisis situations that skills such as workload management, prioritisation and delegation are critical. Rather, even during normal situations it is imperative that individuals and teams arrange work appropriately to maintain safety.

Task management can be defined as the set of organisational activity that is performed as operators initiate, monitor, prioritise and terminate tasks.[1] Together, these activities ensure that optimal working conditions are maintained so that the overarching goals of the work are achieved safely and efficiently. Of critical importance to task management is the optimisation of mental workload to, in turn, ensure optimal work performance.

High-risk industries are typically 'multi-task domains' where an operator must undertake a series of primary and secondary tasks, sometimes in series but often in parallel, to maintain safe and efficient performance. For instance, the control room operator must monitor critical system parameters, communicate with field operators, record actions taken, and manage the inevitable interruptions and distractions. To do this successfully requires aspects of performance such as scheduling and prioritisation of activities, management of their own workload, and also being mindful of work demands placed on others. Together, these skills can be referred to by the term *task management*.

The history of accidents and incidents highlights how task management is a critical domain of non-technical skills. For instance, one study determined that sub-optimal task management caused 23% of commercial aviation accidents and 49% of incidents where safety was significantly compromised.[1]

As with the other domains of non-technical skill, there are various elements of task management that form discrete constituent skills. Indeed, task management is perhaps the domain with the most diverse set of constituent skills. However, each of these skills is relatively less complex and more tangible than the skills underpinning situation awareness and decision-making. This chapter will explore the basic core of these constituent skills, which are as follows:

- Planning and preparing
- Workload management
- Prioritisation
- Delegation
- Managing interruptions

Task Management: Exemplar Case Study

Shortly before midnight on 29 December 1972, an Eastern Airlines Lockheed L-1011 was on approach to Miami airport, completing its regular evening flight EAL 401 from JFK International Airport in New York. As the aircraft made its final approach to Miami, the landing gear was selected down for landing. However, almost immediately, the first officer noticed that the green light to indicate that the nose-wheel was down and locked in position had failed to illuminate. The captain recycled the landing gear lever, but again the green light failed to illuminate.

The crew, identifying that they would not be able to resolve the situation prior to touchdown, called Miami air traffic control and indicated that they would not yet be able to land. Air traffic control cleared flight EAL 401 to climb back up to 2000 feet and to turn north then west, flying away from the airport.

Once the aircraft reached 2000 feet, the autopilot was engaged, and the crew began to troubleshoot the problem. The first officer removed the light assembly, but then it jammed when he was trying to replace it on the instrument panel. While the captain and the first officer discussed repairing the light assembly, the captain requested the flight engineer to go down into the avionics bay below the flight deck and see whether he could visually confirm that the nose-wheel landing gear was locked in the down position.

While the crew were focussing on the problem, somehow the autopilot disconnected, most likely because one of the crew members unwittingly placed pressure on their control wheel. Due to their focus on the landing gear problem, no one heard the altitude alert tone to indicate that the aircraft was descending. Quietly and smoothly, the aircraft slowly descended towards the Florida everglades. Only at the very last minute, and at this stage too late to recover the aircraft, did the flight crew realise they had lost altitude. The aircraft crashed at 2342, only 10 minutes after the crew had first identified the problem with the landing gear light. A total of 101 passengers and crew perished as a result of the accident. Remarkably, 75 survived.

The National Transport Safety Board report into the accident identified the probable cause as the flight crew's failure to monitor the flight instruments in the final four minutes of the flight, allowing the descent to go unnoticed and unchecked.[2]

The trajectory of this tragic accident began with the most simple of problems, a faulty light. However, sub-optimal task management led to all crew members becoming pre-occupied with the light, and no one was left to fly the aeroplane.

Task Management: Core Enabling Knowledge

The first element of core enabling knowledge is to establish the relevance and importance of task management as a domain of non-technical skills and to illustrate its role in contributing to adverse events. This introduction will touch on each of the elements of task management as relevant to the trainees' role and industry context.

The Role of Planning in Performance

Knowledge relating to the critical role of planning in subsequent performance is probably the first element of core enabling knowledge to be introduced. Strong links have been established between planning and team performance across high-risk industries. Research has shown that teams who plan effectively maintain better shared mental models, communicate better and have improved team performance.[3]

There is also a clear link between planning and the non-technical domain of communication. It has been shown that the most effective teams in high-risk industries discuss problems in greater depth and use low-workload periods to discuss options and to plan ahead.[4] Likewise, sub-optimal planning performance is associated with a higher risk of accident or incident, with, for instance, over 20% of medical malpractice incidents being attributed to deficiencies in planning in one study alone.[5]

Core enabling knowledge relating to planning should focus on the specific actions that constitute effective planning, such as the processes of setting goals, assigning roles and responsibilities, discussing operational and environmental constraints and setting expectations for information sharing and problem solving. Planning can occur prior to a specific task, or during a task when a dynamic need arises. Indeed, in high-risk industries, the ability of teams to reconfigure a plan when operational complexities demand it has been demonstrated to be critical for maintaining safety.[6]

Human Information Processing

Although often somewhat complex, a basic understanding of human information processing and its implications for workload and task management is important enabling knowledge. Much research over the last 50 years has attempted to explain how we are able to perform more than one task concurrently, and what our abilities and limitations are with respect to maintaining performance over several concurrent tasks.

Critical to this aspect of human performance is the concept of multiple resources and our ability to maintain secondary task performance, especially when the concurrent tasks use different modalities (visual vs. auditory inputs, or verbal vs. motor skill outputs). To this end, Wicken's 4-D multiple resource model of information processing is a particularly useful tool.[7]

However, just as important as the multiple resource model of concurrent task performance and task switching is an understanding of our limitations with respect to multitasking and our limitations with respect to workload.

Workload Limitations

Another part of basic knowledge associated with effective task management involves developing an understanding of our basic human performance limitations relating to workload. The common way of introducing this concept is through the Yerkes Dodson 'inverted U' theory of performance. This model suggests that performance is best at an optimum level of arousal; low levels of arousal can be described as underload and performance is sub-optimal, while at high levels of arousal we enter a state of overload and again, performance is sub-optimal.[8] However, this model is now rather dated, and alternative models may well be more appropriate for use in high-risk industries.

At the least, an understanding of the main determinants of workload, including (1) personal experience and expertise, (2) task complexity and difficulty, (3) time demands and (4) multiple competing demands, is a critical element of core enabling knowledge.

Indicators of Adverse Workload Conditions

The ability to identify and respond effectively to adverse workload conditions, including both underload and overload, is a critical skill in the task management domain. Understanding the processes that are used to identify adverse workload conditions is therefore core enabling knowledge. Understanding personal workload involves some degree of metacognitive activity, whereby an operator must monitor not only their work environment but also their own mental state. To facilitate this process, a set of 'red flags' for adverse workload conditions has been identified in the literature, and an understanding of these forms is important knowledge. A set of example 'red flags' for adverse workload conditions is provided in Table 10.1.

Table 10.1 Selected Red Flags for Adverse Workload Conditions

UNDERLOAD	OVERLOAD
• Easily distracted/wandering mind	• Stress/anxiety
• Boredom	• Missing steps in a task
• Complacency	• Forgetting certain tasks/task shedding
• Micro-sleeps	• Narrowing focus/tunnelling attention

Compensatory Control and the Effort Monitor

The significant body of research into human performance has suggested that a process of *compensatory control* optimises our task performance. That is to say, the allocation of our resources, and in particular our cognitive resources, is managed in response to task demands. Two mechanisms have been put forward as critical to this process of compensatory control. First, an effort monitor automatically assesses the level of effort we are having to exert in maintaining task performance. Second, a supervisory controller responds to the output of the effort monitor and implements different modes of performance–cost trade-off.[9] Thankfully, these processes are largely automatic in nature! However, what is meant by the performance–cost trade-off is that each task is examined, the elements of the task that have the greatest importance to maintaining safe performance are focussed on, and tasks that have a low cost in terms of potential risk are delayed or abandoned. That is to say, we end up focussing on the critical elements of the task at the expense of less important activities.

Delegation and Distributed Team Performance

The ways in which tasks can be shared between a team, or delegated to individuals, forms another important element of task management. A classic example of this relates to who assumes the role of pilot flying (being actively in control of the aircraft) in a non-normal or emergency situation. This has been discussed and debated over many years, with alternative points of view being regularly put forward. One school of thought suggests that the junior crew member, the first officer, should be given the role of pilot flying, thus freeing up the captain to engage in more immersive higher-order cognitive functions such as problem identification, solution generation, risk analysis and other aspects of decision-making.

Another example of this involves the point at which an anaesthetist will call for assistance when a problem with the patient or equipment arises. This is often a point of significant discussion in debriefing of anaesthesia crisis resource management sessions. Enrolling the assistance of juniors and shedding simple tasks to relieve workload, and enrolling the assistance of seniors to provide a second opinion or assist with greater knowledge and skills, are two strategies that are equally appropriate in many situations. The important thing is to assist trainees in knowing what options exist with respect to delegation and/or assistance, and when different solutions might be appropriate.

Prioritisation of Tasks

Along with delegation and sharing tasks, another important element of task management in high-risk industries involves the prioritisation of tasks and ensuring that the most safety-critical tasks are performed effectively.

In the aviation industry, there has been a long-established axiom of '*aviate, navigate then communicate*'. This provides a powerfully simple mnemonic for task prioritisation, with the first priority being to keep the aircraft in the air, then pointing in an optimal direction, and then to communicate with external parties such as air traffic control. This approach can be easily adapted to a range of other work environments.

The Negative Impact of Interruptions

Interruptions are well known to create an insidious barrier to effective task management in many high-risk industries. Given the limits of our attentional system and our limited ability to perform concurrent tasks, as discussed in the *Compensatory Control and the Effort Monitor* section, any interruption has the potential to lead to omitted tasks and other forms of error. For instance, in a study of approximately 40,000 medication errors, it was found that distractions were identified as a contributing factor in 49% of medication errors.[10]

Research has demonstrated, though, that the effects of interruptions and distractions can be mitigated through effective task management practices. For instance, in a simulator study, anaesthetists who deferred or blocked an interruption made significantly fewer errors in the task at hand than those who engaged with the interruption.[11] This highlights that not allowing an interruption to disrupt current task completion is an effective strategy. However, this is not always an appropriate strategy. Ensuring the correct resumption of a task after it has been interrupted is the key to avoiding errors. Starting the task again from the beginning, or making a note of where you were up to in a task such that items are not omitted, is known to reduce the likelihood of an error resulting from the interruption.

Task Management: Skill Development

A number of specific skills can be developed to assist operators with enhancing task management and avoiding situations where error-producing conditions such as cognitive overload can result in compromised safety.

Leadership

Leadership is an important non-technical skill in and of itself, and the scope of this book has been set such that, unfortunately, it will not be the focus of a dedicated chapter, which, of course, it could easily have been. However, considerable common knowledge and general principles exist with respect to the skills associated with effective leadership and the ways in which they can be developed. Leadership is a critical component of task management in team environments and enables functions such as direction and delegation, which in turn can optimise task scheduling and workload.

Effective Planning Techniques

The first skill used to optimise task management is effective planning. In many high-risk industries, a formalised process governs planning. This ritual of pre-task planning, such as a job safety analysis or 'take-five', has, through its repetitive nature, the potential to lose value over time. To maintain engagement in the planning process, a number of techniques can be used. First, ensuring that all team members actively contribute is an essential skill for team leaders, supervisors and seniors to develop.

Second, the concept of 'threat and error management' has been used to ensure that the planning stage involves active consideration of the job ahead. Having to actively identify both what the potential threats are in the context of the operation, and what potential errors might be made, ensures that active thought goes into the planning process and also primes operators for the risks to the safety and efficiency of the operation.

Recognition of Adverse Workload Situations

The ability to recognise early signs of adverse workload situations is a very important constituent skill to develop. As discussed in the

Indicators of Adverse Workload Conditions section, reinforcing a set of 'red flags' for adverse workload conditions is key to this process.

Appropriate Compensatory Control Strategies to Maintain Safety

As described in the *Compensatory Control and the Effort Monitor* section, a rather complex automatic process governs the way in which we respond to high-workload situations. While the underlying cognitive processes are largely automated, they result in either an increase in cognitive effort to maintain task goals or a change in task goals to reduce cognitive effort. Deploying inappropriate strategies in this situation can result in task overload or a change in task goals that compromises safety.

Therefore, the use of *appropriate* compensatory control strategies should form the target of skill development. One of the most critical aspects of workload and task management through compensatory control involves the strategic scheduling of tasks according to their priority. Many studies of pilot workload have identified a number of general principles to enhance task management. First, ensuring that low-priority tasks are performed during periods of low workload can have a beneficial flow-on effect when workload subsequently increases. Second, being able to identify the high-priority tasks and ensure that these are completed in a timely manner, whereas low-priority tasks are shed, is also critical to maintain escalating levels of workload.[12]

Managing Interruptions

Another critical skill relating to task management involves rehearsing the techniques that can be used to effectively manage interruptions. Strategies discussed previously, such as ignoring, blocking and deferring interruptions, can be the subject of deliberate practice. Similarly, the techniques used to safely resume a task after it has been interrupted are another important focus for skill development.

Prospective Memory Aids

For a human factors practitioner, one of the most fascinating aspects of task management in domains such as aviation and medicine is the informal strategies used by experienced operators to manage tasks and the memory aids they use to support task performance. Many of these strategies involve repurposing aspects of the work environment to serve as reminders for as yet uncompleted tasks. A classic, and unique, example of this is the use of a seat adjustment handle as a reminder that a take-off or landing clearance still needs to be obtained. If the handle is down, the clearance is still needed, and only once the clearance has been obtained is the handle stowed.

Another example is from the medical domain. When a doctor needs to complete a request for a blood test for a patient, they take one of the sticky labels with the patient's identification details and stick it to their sleeve. Then, once they have completed tasks on the ward, they find a computer terminal where they can use the sticky labels as reminders to complete the electronic requests for blood tests. While these techniques are often highly individualised, they can certainly be the source of discussion during non-technical skills training programs.

Task Management: Simulation-Based Training

With respect to learning in general, there is a long-standing understanding that high workload, and in particular cognitive overload, has an overall adverse effect on learning. As trainees expend increasing levels of mental effort to maintain task performance, the efficiency of learning declines.[13] Therefore, there is an optimal level of workload in general for the development of skills. However, training of task management strategies by definition must include responding to instances of increasing workload and deploying strategies to avoid or recover from cognitive overload. This implies that the training of this domain of non-technical skills must be staggered with the development of underlying task-related technical knowledge and skill, such that effective task performance can be maintained, and negative impacts on overall learning can be avoided.

High-Workload Non-Normal Situations

The most obvious mechanism to develop skills in task management in the simulator is through the development of scenarios that create high-workload non-normal situations. These scenarios should be crafted such that they do not create such a high workload that trainees are likely to fail. Rather, the scenarios should be designed to increase workload to a level that can be effectively managed through the deployment of non-technical skills with respect to strategies involving task prioritisation, task shedding, delegation and teamwork.

In the aviation environment, these scenarios typically involve introducing successive aircraft system malfunctions, sometimes in the context of deteriorating environmental conditions. The trainee's threshold for work overload will be different as a function of previous experience and competence. Therefore, rather than running a single scripted scenario, the instructor can have prepared a range of events to introduce sequentially so as to maintain high workload to a degree that still remains manageable. A similar approach can be adopted in any high-risk domain, and is also currently used in anaesthesia with scenarios that overlay equipment problems over a continually deteriorating patient.

The instructor should monitor for signs of overload, such as task fixation, tunnelling and non-verbal cues, such that a high level of workload is maintained but the trainee is not overloaded to the point of severely compromised performance. Similarly, these scenarios test the other dimensions of task management, such as task shedding, delegation and prioritisation of tasks.

Stress Exposure Training

In a similar fashion to the use of high-workload events in simulation-based training, a body of work has reinforced the benefits associated with exposure to stressful events. The objective of this type of training, termed *stress exposure training*, is to assist trainees in developing techniques to maintain adequate task performance in stressful situations, where time pressure, multiple task demands, and environmental stressors such as noise and vibration may degrade performance.[14] Combined with the training of specific cognitive and physiological control strategies, stress exposure training also emphasises the benefits of being desensitised to stressful situations by gradual exposure to increasing levels of external stressors.

Exposure to and Practice Recovering from Unsafe System States

Another approach to the training of recovery from high-workload emergency situations has emphasised the important role of allowing trainees to experience unsafe system states and rehearse both technical and non-technical aspects of recovery. This has specifically been used in aviation with advance jet upset recovery training programs. For instance, the Royal Australian Air Force developed an error management training program, which focussed on the strategies used to identify and recover from situations when the aircraft has transitioned to outside the boundaries of safe operation.[15] Such exposure-based training can assist in reducing the likelihood of cognitive overload and also reduce the negative effects associated with the startle response.[16]

Rehearsal and Extension

Workload management also can be effectively trained through the progressive exposure to more and more demanding situations. As the trainee develops greater proficiency in certain tasks, additional competing tasks can be introduced, or time pressure can be increased.

Task Management: On-the-Job Training

A supervisor or coach can use a number of techniques to support the development of task management skills on the job. Indeed, much of what a supervisor does is supporting the task management processes of teams in high-risk environments.

Pre-Work Planning

The first critical element of training task management skills on the job is to facilitate effective pre-task planning. Most high-risk industries have pre-task planning as a formal requirement, and it is often supported with risk assessment and job planning aids. Ensuring that these processes are followed in a manner that ensures participant engagement is an important role of the supervisor or coach.

Contingency Planning

Another aspect of task management that lends itself perfectly to on-the-job training is contingency planning. Even as work unfolds in the dynamic environment, trainees can be asked to pause and reflect on what would be the best course of action or solution if a specific event were to occur. Subsequent coaching can reinforce the optimal technical and non-technical responses.

This process of posing a series of 'what if?' scenarios is particularly useful in low-workload stages of a task, such as during in-flight cruise, after induction of anaesthesia, or after any natural pause in the proceedings of work. It maintains cognitive engagement with the task, primes trainees in the case of that scenario actually occurring, and can activate important technical and procedural knowledge to maintain its recency and thus prevent decay.

Post-Event Debrief

A final strategy for developing task management strategies on the job is to facilitate critical reflection after a complex, high-workload event has been encountered. Drawing on the theory of professional development and reflective practice, non-technical skills such as task management can be constantly refined by critical reflection.

Task Management: Pre-Training Briefing

Pre-training briefing for non-technical skill development for task management should focus on activating core enabling knowledge and then exploring strategies for effective task management relevant to the training scenario at hand. Priming trainees for a variety of potential situations by way of prompting contingency planning is also a useful technique.

Allowing Training Participants to Plan and Brief Themselves

In normal work environments, pre-task planning is a key to effective task management. This should not be any different in the training environment, especially prior to authentic simulation-based or real-world training events. Allowing trainees to properly prepare themselves with their own planning and briefing activities will enable effective task management from the outset. Therefore, once the instructor has facilitated the activation of core enabling knowledge and discussed core task management strategies, s/he should then allow time and space for the trainees to prepare themselves.

AN EXAMPLE IN PRACTICE

The track maintenance team in the railway environment face considerable challenges with respect to task management, as their work often needs to be coordinated around the continual movement of trains along a section of track.

In light of a series of incidents, it was decided that a coaching program for track maintenance teams would be developed to further enhance their task management and coordination skills. Coaches were trained in the essential elements of task management and given tools for the observation of teams on track.

A critical component of this coaching activity was to ensure that track maintenance teams performed the highest-quality briefing prior to commencing work. Additional time was set aside at the commencement of work for the day for the track maintenance coach to brief in detail how the work was to be performed, ensure that roles and responsibilities were clear in everybody's minds, and perform a detailed risk assessment.

Task Management: Assessment and Behavioural Markers

The domain of task management includes constituent skills that can be easily observed and assessed, as well as skills that are less easy to observe. For instance, workload management strategies can only be inferred through observable behaviours such as task shedding, prioritisation and delegation. Accordingly, assessing task management performance requires a combination of self-report techniques and behavioural markers.

Self-Report Workload Scales

As mental workload is to a large degree an individual and subjective phenomenon, it lends itself to the use of self-report scales as part of the assessment process. A range of tools can be adapted from the research literature for use in assessing performance during training. The NASA Task Load Index (TLX) is one example of such a self-assessment tool that is able to be quickly and efficiently deployed in the training environment.[17,18] The tool prompts trainees to rate the workload on six scales: (1) mental demand; (2) physical demand; (3) temporal demand; (4) effort; (5) performance; and (6) frustration. The tool can be quickly administered, can provide both instructor and trainee with considerable insights into perceived workload, and can be used as an effective prompt for facilitated debrief of the training session. Once the level of workload has been assessed, discussion can focus on the strategies used to manage either very low or very high levels of workload. A range of other self-report scales can be used for measuring workload, and some of these are summarised in Table 10.2.

Behavioural Markers

Apart from workload, the other constituent skills associated with task management are readily observable through the overt behaviour of the trainee and therefore lend themselves to assessment using behavioural markers. Like communication, the domain of task management

Table 10.2 Self-Report Workload Scales That Can Be Adapted for Assessing Task Management Performance during Training

TOOL	DESCRIPTION
Subjective workload assessment technique (SWAT)[19]	A simple tool that measures three dimensions: (1) time load; (2) mental load; and (3) stress load. Each of these dimensions is measured on a three-point scale from low to high.
Malvern capacity estimate (MACE)[20]	A simple technique that can be used during a training session. The trainee is simply asked to estimate their remaining workload capacity.
Bedford scale[21]	A decision-tree that prompts trainees to rate workload on 10 levels, from being so high that the task was not able to be completed through to so low that workload was insignificant.

is quite broad, and as such, a range of different elements, or constituent skills, have been identified. Table 10.3 highlights a range of these constituent skills.[22,23]

As there are multiple techniques that can be used for a specific area of task management, and each of these might be appropriate, a different approach to the development of behavioural markers can be adopted. If we take the work on interruptions to anaesthetists discussed in the *Negative Impact of Interruptions* section,[11] it is possible to develop a classification system that prompts the instructor to define and analyse the effectiveness of the task management techniques used. An example of what this might look like in practice is provided in Table 10.4.

Overall, there are a number of ways in which behavioural markers and checklists can be developed to assist the instructor in assessing non-technical performance. The primary thing to keep in mind is that effective support needs to be provided to ensure validity, reliability and sensitivity in assessing non-technical skills.

Task Management: Debriefing and Coaching

A number of different techniques lend themselves to the process of debriefing and coaching task management skills, and each of the techniques discussed in previous chapters can be effectively used.

Table 10.3 Typical Elements within the Domain of Task Management

CATEGORY	ELEMENT
	Planning and preparing
Task management	Workload management
	Prioritising
	Delegation
	Managing stress
	Managing interruptions

Table 10.4 Specific Behavioural Markers for Managing Interruptions

TECHNIQUE	DESCRIPTION
Ignore	Did the trainee ignore the attempted interruption? Was this successful?
Defer	Did the trainee acknowledge the interruption but defer it for later? Was this successful?
Engage	Did the trainee engage with the interruption? Did this impact on performance?
Recovery	What did the trainee do to recover from the interruption to ensure safe task performance? Was this successful?

While video can be useful, the primary aim of debriefing and coaching task management skills should be focussing on whether the strategies used during the training event were appropriate and effective. Accordingly, it is necessary to facilitate debrief with respect to how resources were managed, how tasks were prioritised and how workload impacted on performance.

This type of debrief requires a significant degree of critical self-reflection from trainees. To achieve the reflection, a selection of trigger questions that could prompt facilitated debrief is as follows:

- How well did you think your planning assisted you in the task?
- Was there anything you did not anticipate at the planning stage?
- Do you think your performance was affected by any of the interruptions?
- How did you prioritise tasks when things got busy?
- Do you think your workload impacted on your performance?
- What other strategies could you use to manage the stress of the situation?

Task Management: Key Resources

The following are excellent resources that explore in detail elements of task management as they relate to high-risk work domains.

Flin, R. H., O'Connor, P., and Crichton, M. (2008). *Safety at the Sharp End: A guide to Non-Technical Skills*. Aldershot, UK: Ashgate Publishing.

Gawande, A. (2010). *The Checklist Manifesto*. London, UK: Profile Books.

Hancock, P. A., and Desmond, P. A. (2001). *Stress, Workload, and Fatigue*. Mahwah, NJ: Lawrence Erlbaum Associates.

Case Studies

Amtrak Derailment in Philadelphia: On 12 May 2015, an Amtrak passenger train entered a curve at over twice the required speed. The curve had a speed limit of 50 mph, and the train was travelling at 106 mph. Due to the excessive speed, the train derailed. As a consequence, 8 passengers died, and 185 others were injured. The National Transport Safety Board (NTSB) investigation determined that the driver was distracted by radio conversations between a train dispatcher and another locomotive engineer about an emergency situation on another train and failed to decelerate for the upcoming curve.[24]

Delta Airlines Flight 1141: On 31 August 1988, the crew of a Delta Air Lines Boeing B727 was preparing for take-off from Dallas Fort Worth International Airport in Texas. After pushback from the gate, considerable casual and non-operational conversation took place among the flight crew, including with a member of the cabin crew who entered the flight deck on several occasions. The NTSB found that due to excessive non-operational conversation and poor checklist discipline, the crew failed to extend the aircraft's flaps or slats into the take-off position. As a consequence, the aircraft was unable to climb and crashed off the end of the runway, leading to the deaths of 14 passengers and crew.[25]

References

1. Chou, C.-C., Madhavan, D., and Funk, K. (1996). Studies of cockpit task management errors. *The International Journal of Aviation Psychology*, *6*(4), 307–320.
2. NTSB. (1973). *Eastern Airlines Inc L-1011 N310EA Miami Florida December 29 1972 (NTSB-AAR-73-14).* Washington, DC: National Transport Safety Board.
3. Stout, R. J., Cannon-Bowers, J. A., Salas, E., and Milanovich, D. M. (1999). Planning, shared mental models, and coordinated performance: An empirical link is established. *Human Factors*, *41*(1), 61–71.
4. Orasanu, J. M., and Fischer, U. (1992). Team cognition in the cockpit: Linguistic control of shared problem solving. In *Proceedings of the 14th Annual Conference of the Cognitive Science Society* (pp. 189–194). Hillsdale, NJ: Erlbaum.
5. Risser, D. T., Rice, M. M., Salisbury, M. L., Simon, R., Jay, G. D., Berns, S. D., and Consortium, M. R. (1999). The potential for improved teamwork to reduce medical errors in the emergency department. *Annals of Emergency Medicine*, *34*(3), 373–383.
6. Boy, G. A., and Schmitt, K. A. (2013). Design for safety: A cognitive engineering approach to the control and management of nuclear power plants. *Annals of Nuclear Energy*, *52*, 125–136.
7. Wickens, C. D. (2008). Multiple resources and mental workload. *Human Factors*, *50*(3), 449–455.
8. Yerkes, R. M., and Dodson, J. D. (1908). The relation of strength of stimulus to rapidity of habit-formation. *Journal of Comparative Neurology and Psychology*, *18*(5), 459–482.
9. Hockey, G. R. J. (1997). Compensatory control in the regulation of human performance under stress and high workload: A cognitive-energetical framework. *Biological Psychology*, *45*(1), 73–93.

10. Santell, J. P., Hicks, R. W., McMeekin, J., and Cousins, D. D. (2003). Medication errors: Experience of the United States Pharmacopeia (USP) MEDMARX reporting system. *The Journal of Clinical Pharmacology*, *43*(7), 760–767.
11. Liu, D., Grundgeiger, T., Sanderson, P. M., Jenkins, S. A., and Leane, T. A. (2009). Interruptions and blood transfusion checks: Lessons from the simulated operating room. *Anesthesia & Analgesia*, *108*(1), 219–222.
12. Raby, M., and Wickens, C. D. (1994). Strategic workload management and decision biases in aviation. *International Journal of Aviation Psychology*, *4*(3), 211.
13. Paas, F. G., and Van Merriënboer, J. J. (1993). The efficiency of instructional conditions: An approach to combine mental effort and performance measures. *Human Factors: The Journal of the Human Factors and Ergonomics Society*, *35*(4), 737–743.
14. Driskell, J. E., and Johnston, J. H. (1998). Stress exposure training. In J. A. Cannon-Bowers and E. Salas (Eds.), *Making Decisions under Stress: Implications for Individual and Team Training* (pp. 191–217). Washington, DC: American Psychological Association.
15. Naikar, N., and Saunders, A. (2003). Crossing the boundaries of safe operation: An approach for training technical skills in error management. *Cognition, Technology and Work*, *5*, 171–180.
16. Martin, W. L., Murray, P. S., Bates, P. R., and Lee, P. S. (2015). Fear-potentiated startle: A review from an aviation perspective. *The International Journal of Aviation Psychology*, *25*(2), 97–107.
17. Hart, S. G., and Staveland, L. E. (1988). Development of NASA-TLX (task load index): Results of empirical and theoretical research. *Advances in Psychology*, *52*, 139–183.
18. NASA. (1987). *NASA Task Load Index (NASA TLX)*. Moffett Field, CA: NASA Ames Research Center.
19. Reid, G. B., Potter, S. S., and Bressler, J. (1989). *Subjective Workload Assessment Technique (SWAT): A User's Guide*. Wright Patterson Air Force Base, OH: Harry G. Armstrong Aerospace Medical Research Laboratory.
20. Goillau, P. J., and Kelly, C. J. (1997). Malvern capacity estimate (MACE): A proposed cognitive measure for complex systems. In D. Harris (Ed.), *Engineering Psychology and Cognitive Ergonomics* (pp. 219–225). Aldershot, UK: Ashgate Publishing.
21. Roscoe, A. H., and Ellis, G. A. A subjective rating scale for assessing pilot workload in flight: Technical report TR90019. Farnborough, UK: Royal Aerospace Establishment – Ministry of Defence.
22. Fletcher, G., Flin, R., McGeorge, P., Glavin, R. J., Maran, N. J., and Patey, R. (2003). Anaesthetists' non-technical skills (ANTS): Evaluation of a behavioural marker system. *British Journal of Anaesthesia*, *90*(5), 580–588.
23. Klampfer, B., Flin, R., Helmreich, R. L., Hä usler, R., Sexton, B., Fletcher, G., Field, P., et al. (2001). *Enhancing Performance in High Risk Environments: Recommendations for the Use of Behavioural Markers*. Ladenburg: Daimler-Benz Stiftung.

24. NTSB. (2016). *Derailment of Amtrak Passenger Train 188 Philadelphia, Pennsylvania May 12, 2015*. Washington, DC: National Transport Safety Board.
25. NTSB. (1989). *Delta Air Lines, Inc. Boeing 727-232, N473DA, Dallas-Fort Worth International Airport, Texas August 31, 1988*. Washington, DC: National Transport Safety Board.

11

The Future of Training and Assessing Non-Technical Skills

Summary

This book has set out to provide a practical guide to training and assessing non-technical skills. While it has been written with the range of high-risk industries in mind, it is hoped that much will also have been relevant to any modern and future workplace.

The first part of the book has set out a brief history of training and assessing non-technical skills, and emphasised the context that most of what we see in the focus and structure of our non-technical skills training programs in high-risk industries is based on the original development of crew resource management programs in commercial aviation.

The second part of this book has explored in some detail our understanding of how adults learn and the unique approach they bring, which must be accommodated in the design of non-technical skills training programs. This part of the book then introduced the general principles of training and assessment as they relate to non-technical skills and concluded with an overview of instructional design processes that can be used to build an effective program for training and assessing non-technical skills.

The final part of this book explored specific details as they relate to each of four major domains of non-technical skills: (1) situation awareness; (2) decision-making; (3) communication; and (4) task management. These are but four of the main domains of non-technical skills, and there are obviously others that this book did not touch on.

This chapter sets out some thoughts on the future of programs that set out to train and assess the non-technical skills of operators in high-risk industries. There is much we can continue to do to develop these training programs and create even more effective ways to enhance performance and safety.

A More Integrated Approach to Skill Development

One area in which we must continue to refine our approach relates to the false divides that have historically been created in the overall training and assessment of non-technical skills.

The first of these divides, which was introduced early in the book, is that the division between non-technical and technical skills is false and potentially misleading. Take, for example, the process of decision-making in a critical situation. An optimal decision cannot be made by someone who is simply well versed in decision-making theory and technique. Expert domain knowledge and technical skills are also critical to an effective decision outcome. Indeed, as we saw in the decision-making chapter, decision-making skills cannot be trained and assessed in an environment abstracted from real-world workplace situations.

The second false divide relates to the domains of non-technical skills themselves. Again, if we take the example of decision-making by a team in a crisis situation, the outcome cannot possibly be optimal without both situation awareness and communication. The domains of non-technical skills described in this book have been created to conveniently label aspects of expert performance. When aspects of performance are named in such a way, they can be studied discretely and trained in a targeted fashion. However, this approach is highly reductionist, and while it is convenient, non-technical skills training programs must attempt to integrate and bridge this divide.

Our future development of non-technical skills training programs must focus more keenly on the ways in which they can be successfully integrated with aspects of technical training and on-going professional development.

A More Targeted Approach to Skill Development

Currently, many of the non-technical skills training programs seen across high-risk industries attempt to train a full spectrum of non-technical skills within a single training program. This approach has been born in part from the novelty of the need to train non-technical skills, and in part from the limitations of time and money that can be invested in on-going professional development. However, this approach results in many non-technical skills programs covering many topics and areas of non-technical skills but with not a lot of detail.

This book has emphasised that a domain such as situation awareness or task management describes a very broad area of performance, and in reality, these domains are made up of many constituent skills. With this in mind, the current attempt to develop training programs to cover all areas of non-technical skills cannot be seen to truly contribute to detailed skill development. This is especially the case where we have seen the need for deliberate practice as a crucial aspect of skill development.

To this end, future non-technical skills training programs should seek to be much more narrow in their focus and integrate more opportunities for deliberate practice in each of the constituent skills that make up a domain of non-technical skill.

Harnessing the Potential of New Technologies

A number of new and emerging technologies have significant potential for enhancing training and assessing non-technical skills. Current forms include digital recording of audio and video from multiple camera angles, and presented alongside high-fidelity representations of critical operational parameters, these will only continue to advance. These technologies to assist in the process of debriefing and assessment are already probably under-used, and new work is needed to explore how these technologies can be harnessed to maximise their benefits for the learning process.

One of these technologies with considerable potential is that of virtual reality training environments. Virtual reality has been used for a considerable time in the training of technical skills, including skills as diverse as welding and laparoscopic surgery. However, with recent technological advancements, it is now possible to create virtual reality immersive worlds whereby remote individuals can interact in a virtual space. This provides opportunities to create high-fidelity simulation environments without the expensive investment in simulation hardware.

The introduction of these new technologies into non-technical skills training programs demands that they are subjected to scientific evaluation in terms of their effectiveness in enhancing learning. While we are naturally attracted to new technologies, we need to be sure that they do not inadvertently impact on the desired processes of knowledge, skill and attitude development.

Building More Responsive Training Systems

This book began with the premise that training and assessing non-technical skills is a fundamental component of enhancing safety and efficiency in high-risk industries. The history of catastrophic failures in these industries is also the history of failures of individual and team performance.[1,2] Non-technical skills training programs emerged from within this context as an important element of the management of organisational risk. Therefore, as part of an organisation's risk management system, non-technical skills training programs need to be seen in terms of a *responsive* training system.

Throughout this book, we have explored different sources of information that can be used for the development of many aspects of a non-technical skills training program, from training needs analysis, to the choice of appropriate case studies, through to the development of behavioural marker systems. By the term *responsive training system*, we mean that an organisation uses its own sources of data from within its safety management system (SMS) to ensure that the non-technical skills training program identifies and responds effectively to organisational needs. To this end, the non-technical skills training program becomes a critical component of the SMS and the overarching risk management actions of the organisation. This relationship is critical in high-risk industries if we are to achieve the aim of the non-technical skills training program contributing directly to enhanced organisational performance.

Training across Distributed Systems

Early in the history of non-technical skills training in aviation, the need to move beyond the isolated focus on the flight deck to include cabin crew was identified as critical to enhancing training. In any high-risk industry, safe and efficient operations are created as an emergent feature of complex distributed systems. It therefore makes sense to focus skill development in the context of these distributed systems, rather than just focussing on the individual operators in isolation.

True multidisciplinary training of non-technical skills has come up against a considerable number of barriers: issues relating to 'tribal' differences in attitudes between professional domains, hierarchy and organisational silos, or even working for completely different organisations.

However, in reality, each of these barriers reflects issues that are also seen as contributory factors to incidents and accidents in these same industries. For instance, breakdowns in coordination between control room operators and field operators are often cited as an area where non-technical deficiencies have resulted in safety being compromised. However, the barriers highlighted often prevent these groups coming together for multidisciplinary training programs.

To this end, the future of non-technical skills training needs to continue to challenge those barriers and reflect the nature of distributed multi-discipline work in training.

A Final Note

Personally, it has been a privilege to work in the area of non-technical skills now for several decades. There is nothing more fascinating and rewarding than to observe experts at work in environments as diverse as the flight deck of a commercial airliner, the operating theatre, the locomotive cab of a 1.5 km long freight train, the control room of an electrical utilities company, and beyond.

In this time, the domain has continued to mature, and an ever-increasing body of research has been built demonstrating the important role of training and assessing non-technical skills. However, we need to keep striving to improve our training programs and develop new techniques for training and assessing non-technical skills. These are not soft skills that are hard to specify and subjective in their assessment. Today, and into the future, non-technical skills training programs will expand into new industrial contexts, and with each new development, our repertoire of training and assessment techniques must also continue to expand.

References

1. Reason, J. (1990). *Human Error*. Cambridge, UK: Cambridge University Press.
2. Reason, J. (1997). *Managing the Risks of Organizational Accidents*. Aldershot, UK: Ashgate Publishing.

Glossary

Throughout this book, we have referred to a large number of key concepts relating to the overall design, development and implementation of non-technical skills training programs. The following glossary provides a brief orientation to these key concepts that are referred to in this book.

adult learning theory: a set of principles that set out the specific characteristics of how adults learn, in comparison to the ways in which learning takes place throughout childhood in formal educational institutions

assessment: the process of measuring a trainee's knowledge, skills or attitudes

attitudes: a set of beliefs that predisposes someone to certain behaviours and actions

awareness training: developing an understanding of the importance of non-technical skills to safe and efficient operations in high-risk industries

behavioural marker: a description, or word picture, that describes the specific behaviours related to a specific aspect of non-technical performance. Typically, a behavioural marker will provide examples of specific actions that describe good and poor performance

bias: a cognitive predisposition to act in a certain manner

briefing: the activity undertaken prior to training in which the trainees are primed for learning. A briefing typically outlines the objectives of the training session, revisits core enabling knowledge and sets the expectations for performance during training

case study: a pre-defined scenario or event presented to trainees in the form of a problem, which the trainees must attempt to solve

coaching: a process of observing work on the job and providing directed feedback with respect to performance and areas for enhancement

communication and teamwork: the domain of non-technical skills that relates to the social interaction of people

competence: the knowledge, skills and attitudes required to perform a task

competency-based training: training that develops the knowledge, skills and attitudes required to perform a task to a defined standard

core enabling knowledge: the knowledge that needs to be acquired to achieve optimal performance in a specific non-technical skill

criterion-referenced assessment: assessment that compares a trainee's performance with a pre-determined criterion. This concept is critical in non-technical skills training programs that are competency-based

curriculum: the set of topics and the learning and assessment activities that together form the whole of a training program

curriculum outcomes: the overall aims and objectives of a training program

debriefing: the activity undertaken after a training event, in which trainees are able to critically reflect on aspects of good and poor performance, and the instructor is able to provide feedback on performance

decision-making: the domain of non-technical skills that relates to the processes of identification that something is not as planned, diagnosis of the situation, and implementation of a suitable choice of action

facilitation: providing assistance to trainees through actions that will help them learn. In non-technical skills training programs, this is typically through assisting in the learning processes of discussing and reflecting on performance, and providing guidance with respect to optimal and sub-optimal strategies

fidelity: the degree to which a simulation reproduces an aspect of real-world operations

formative assessment: assessment of knowledge and/or skill development that takes place during the course of the training program. This is typically used to measure and give feedback on learning progress and to tailor additional training interventions

heuristic: A mental short-cut or 'rule of thumb'

instruction: undertaking a range of actions to facilitate learning, including such things as describing, guiding, facilitating, coaching and assessing

instructional objectives: the desired outcomes of training described in terms of new knowledge, skills or attitudes. Usually derived from formal training needs analysis

instructional systems design: a set of formal processes through which a training program is designed, implemented and evaluated

knowledge: an understanding of concepts and facts

learning: the process of acquiring new knowledge, skills or attitudes, resulting in behaviour change

mastery: another term for *proficiency*, used as a criterion for cessation of training when errorless performance is consistently achieved in a certain task

mnemonic: a memory aid, often presented in the form of an acronym designed to facilitate an optimal sequence of steps or an ideal course of action

non-technical skills: the cognitive and social abilities that complement the technical skills of workers and contribute to safe and efficient performance in high-risk industries. They include competencies within the domains of situation awareness, decision-making, task management, and communication and teamwork

normative-referenced assessment: assessment that compares a trainee's performance with those of their peers. This form of assessment is not appropriate for non-technical skills training programs that are competency-based

on-the-job training: either formal or informal training interventions that take place during normal work situations

proficiency: the ability to do something consistently and with a high degree of safety and efficiency

reliability: the degree to which an assessment method is able to consistently produce the same outcome

role-play: a training activity that involves presenting a hypothetical situation to a group and selecting individuals to assume roles within that scenario, which is then 'acted out' to facilitate skill development

safety management system (SMS): the set of organisational processes for the identification of hazards, mitigation of risk, measurement of performance, investigation of incidents, and on-going continuous improvement

simulation: the artificial re-creation of a real-world work environment to facilitate training

situation awareness: the domain of non-technical skills that relates to the perception of critical information, comprehension of what this information means, and projection of what the current state means for the future

skill: the ability to perform a series of actions consistently and accurately

summative assessment: assessment of knowledge and/or skill development that takes place at the completion of a training program. This is often referred to as a *test*, *examination* or *proficiency check* in high-risk industries

syllabus: the set of topics being covered in a training program

task management: the domain of non-technical skills that relates to how tasks are planned, coordinated and executed effectively to maintain safe and efficient operations in high-risk industries

training: A *formal process* designed to ensure that trainees develop the knowledge, skills and attitudes required for work performance

training intervention: a specific training event, such as a classroom seminar, a simulation-based scenario or a case study designed to promote learning. We will use this term to describe individual units of training

training needs analysis: a process of identifying the specific knowledge, skills and attitudes needed for safe and efficient work performance

validity: the degree to which an assessment method measures the aspect of performance it claims to measure

Index

C

T

W

Printed in the United States
By Bookmasters